Polymer-Based Aqueous-Two Phase Systems

Masami Shibukawa

Polymer-Based Aqueous-Two Phase Systems

Fundamentals and Applications

Springer

Masami Shibukawa
Graduate School of Science
and Engineering
Saitama University
Saitama, Japan

ISBN 978-981-96-2887-2 ISBN 978-981-96-2888-9 (eBook)
https://doi.org/10.1007/978-981-96-2888-9

This Springer imprint is published by the registered company Springer Nature Singapore Pte Ltd.
The registered company address is: 152 Beach Road, #21-01/04 Gateway East, Singapore 189721, Singapore

Preface

An aqueous solution of two water-soluble but mutually incompatible compounds may form two immiscible aqueous phases under certain conditions of concentration, temperature, and pressure. The two-phase system thus formed is called aqueous two-phase system (ATPS) or aqueous biphasic system (ABS) and has been investigated by many researchers as a platform for a liquid–liquid extraction to replace conventional methods using organic solvents.

ATPS can be formed with a variety of components in aqueous solution. Recently, ATPSs with ionic liquids and deep eutectic solvents as phase-forming components have been intensely studied. However, extraction with polymer-based ATPS, especially polyethylene glycol (PEG)-based ATPS, has been the most widely studied and used as an environmentally friendly extraction technology. This book presents the fundamentals of the separation mechanism of the polymer-based ATPS and applications of ATPS for the separation and/or enrichment of inorganic compounds with batch extraction and chromatographic techniques.

The elucidation of the solute partitioning mechanism at the molecular level in these systems is of vital importance for their successful application. However, the extraction mechanism in polymer-based ATPS has not been elucidated until recently. The author first describes the microstructures of the aqueous polymer phases formed in the PEG-based ATPS clarified by equilibrium analysis of a metal complexation reaction in aqueous PEG solutions and differential scanning calorimetry. Then the mechanism of solute partitioning in PEG-based ATPS is discussed based on the structure of the aqueous PEG solution phase. A subsequent chapter shows the applications of the ATPS batch extraction to the analyses of metal ions in environmental samples, i.e., river water, soil, and sediment samples. In the following chapters, the author introduces liquid chromatography (LC) with hydrophilic polymer gel columns using an aqueous salt solution as the mobile phase, which is regarded as an efficient separation and enrichment technique based on polymer-based ATPS.

Lastly, the author describes multistep pH-peak-focusing countercurrent chromatography and LC and partition/ion-exclusion chromatographic ion stacking (PIEC ion stacking), newly developed ATPS chromatographic techniques for separation and enrichment of metal ions and inorganic anions, in detail. ATPS is a unique separation platform that can incorporate hydrophilic and ionic solutes in their native forms in both of the two phases. This book helps researchers and engineers design and implement novel or appropriate environmentally benign analytical methods for environmental analyses, reagent manufacturers, metal refining industry, and so on, by not only a conventional batch-wise extraction method but also a more useful and high-performance liquid chromatographic method.

Saitama, Japan Masami Shibukawa
sibukawa@mail.saitama-u.ac.jp

Acknowledgements I would like to express my appreciation to my colleagues and graduate students at Chiba University, Nihon University, and Saitama University for their technical assistance and valuable discussions on the research works. I am also indebted to my wife, Atsuko, as a whole for her patience and continuous support when undertaking my research and writing this book. Finally, I take this opportunity to express my thanks to Mr. Koizumi of Springer Nature for his tenacious support.

Competing Interests The author has no competing interests to declare that are relevant to the content of this manuscript.

Contents

Chapter 1
Introduction to Extraction and Chromatography with Polymer-Based Aqueous-Two Phase Systems

Abstract Liquid–liquid extraction is one of the most widely used separation methods that exploits the difference in solute partitioning between two immiscible liquid phases and is usually performed using water and water-immiscible organic solvents as its separation mechanism has already been clarified by many studies. On the other hand, aqueous two-phase systems (ATPSs), formed by mixing two water-soluble but mutually incompatible compounds in water, have emerged as an environmentally friendly platform for liquid–liquid extraction. Although ionic liquids (ILs) and deep eutectic solvents (DESs) have been recently introduced as new types of phase-forming components for ATPS, polymer-based ATPS, especially with polyethylene glycol (PEG), is the most widely used platform for liquid–liquid extraction at present in the scope of the green chemistry concept. The author will demonstrate in this book the effectiveness of polymer-based ATPS as chemical separation systems by describing the fundamentals of the separation mechanism of the ATPS and its applications to separation and/or enrichment of inorganic compounds with a conventional batch-wise liquid–liquid extraction technique and newly developed ATPS liquid chromatographic techniques. This chapter provides the fundamentals of polymer-based ATPS and an overview of the contents of this book.

Keywords Aqueous two-phase system · Phase diagram · Liquid–liquid extraction · Solute distribution mechanism · Liquid chromatographic techniques with aqueous two-phase systems

1.1 Background

When two water-soluble but mutually incompatible compounds are dissolved in water, the aqueous solution may separate into two phases above critical concentrations at given pressure and temperature conditions. In such a two-phase system, one compound is found predominately in one phase and the other one in the other phase. The two-phase system thus formed is called aqueous two-phase system (ATPS) or aqueous biphasic system (ABS). ATPS was first discovered by Beijerinck in 1896

M. Shibukawa, *Polymer-Based Aqueous-Two Phase Systems*,
https://doi.org/10.1007/978-981-96-2888-9_1

while mixing aqueous starch and gelatin solutions or aqueous agar and gelatin solutions [1, 2]. However, it was much later that the potential of ATPS was recognized by Albertsson as effective separation and purification systems for biological materials such as proteins, cells, and viruses [2, 3].

Liquid–liquid extraction is a method of separating substances using the phenomenon of solute partitioning between two immiscible liquid phases, usually water and an immiscible organic solvent, and is used as an essential technique in many industrial processes and chemical analyses. Separation can be achieved by mass transfer of target solute compounds from one liquid phase to another, generally from aqueous phase to organic phase using relatively simple apparatuses such as separatory funnels. However, a number of organic solvents employed for conventional liquid–liquid extraction are toxic, flammable, and volatile, which may contribute to increasing the risks of fire and explosion and can act as environmental pollutants such as carcinogens and greenhouse gases [4]. In addition, hydrophilic molecules and ions are poorly extracted to water-immiscible organic solvents, so suitable derivatization is usually needed to reduce their hydrophilicity and enhance the extractability. Liquid–liquid extraction with water-immiscible organic solvents cannot also be applied to the purification of biological compounds because most organic solvents denature biomolecules.

The most conventional ATPSs are polymer-based ones and are formed by mixing a water-soluble polymer with a dissimilar kind of polymer or with an inorganic salt in water. Among a lot of kinds of polymers that have been investigated so far, polyethylene glycol (PEG) is most commonly used as a phase-forming polymer for ATPS because it is commercially available, inexpensive, nonflammable, and nontoxic. The toxicity for a wide molecular range of PEGs has been assessed in detail and concluded to be safe for consumption [5–8]. Therefore, in contrast to the traditional liquid–liquid extraction system, ATPSs may be thought of as environmentally benign media for extraction that is directly applicable to the separation of hydrophilic solutes including biomolecules and charged compounds since they are mainly composed of water and phase-forming components are virtually nontoxic [2, 9, 10].

PEG/dextran and PEG/salt ATPSs have been widely used for separation of not only biological materials but also inorganic compounds [9–15], where the PEG-rich phase usually acts as an extraction medium exhibiting moderate hydrophobicity. PEG/dextran ATPS is preferable for the separation of biological molecules which undergo denaturation or degradation at high ionic strength [11]. However, they usually exhibit small differences in polarities between the two phases resulting in restricted separation selectivity [16]. On the other hand, PEG/salt ATPS can present much larger separation selectivity for a wide range of solutes including inorganic compounds such as metal ions as well as organic compounds [11, 16]. The disadvantage of polymer-based ATPS, which requires a relatively long time for phase separation due to their relatively large viscosity, is considerably mitigated in PEG/salt ATPS compared to PEG/dextran ATPS.

In recent years, ATPS composed of ionic liquids (ILs) has been explored to greater extent to extend the range of applicability of the extraction technique with ATPS. ILs are salts which have melting temperatures below 100°C, originating from the combination of large organic cations and inorganic or organic anion [17]. IL-based ATPS is characterized by the ability to control extraction selectivity relatively easily by changing the structures of ions and by rapid phase separation compared with polymer-based ATPS in addition to their high biocompatibility and low volatility. However, some ILs have the disadvantage that they are not entirely nontoxic although water-miscible ILs used to prepare ATPS are less toxic than hydrophobic-based counterparts [17, 18]. The synthesis and purification of ILs are also usually complex [19, 20].

In addition to ILs, deep eutectic solvents (DESs) have been recently emerged as a new type of phase-forming components for ATPS [19–22]. DESs are liquids formed by complexation through hydrogen bonding between hydrogen bond acceptor such as choline chloride and betaine and hydrogen bond donor such as glycol and glycerin. DESs are considered to be biodegradable and less toxic compared to ILs [19, 20] and DES-based ATPS has the potential to become another sustainable platform of extraction.

Although various types of compounds have been proposed as phase-forming components for ATPS as described above, polymer-based ATPS is the most widely investigated and used eco-friendly platform for liquid–liquid extraction in the scope of the green chemistry concept. This book is intended to demonstrate the effectiveness of polymer-based ATPSs as chemical separation systems by describing the fundamentals of the separation mechanism and applications of the ATPSs to separation and/or enrichment of inorganic compounds with batch-wise liquid–liquid extraction and chromatography techniques.

1.2 Phase Diagram

As the ATPS consists of three components, water and two water-soluble compounds, the relative proportions of the three components in the system can be represented by ternary phase diagrams or triangular diagrams. A typical ternary phase diagram for the polymer/salt ATPS composed of PEG6000 (average molecular weight = 6000)/Na_2SO_4/water is shown in Fig. 1.1a, in which the composition is given by the percentage of weight fraction. However, in order to define the composition of a ternary system, it is necessary to specify only the fractions of two components, since the fraction of the third component can be obtained from those of the other two components [23, 24]. Thus, an orthogonal or rectangular phase diagram that expresses the weight fractions of the components in the two phases other than water is usually used (Fig. 1.1b).

In the orthogonal phase diagram shown in Fig. 1.1b, the binodal line separates the phase diagram into the homogeneous (single-phase) and heterogeneous (two-phase) regions. The three component mixtures whose total compositions are above the binodal line give two immiscible phases, while those whose total compositions are below the line present a single phase. For a given total mixture composition in the two-phase region (T), the mixture is phase separated and forms two coexisting

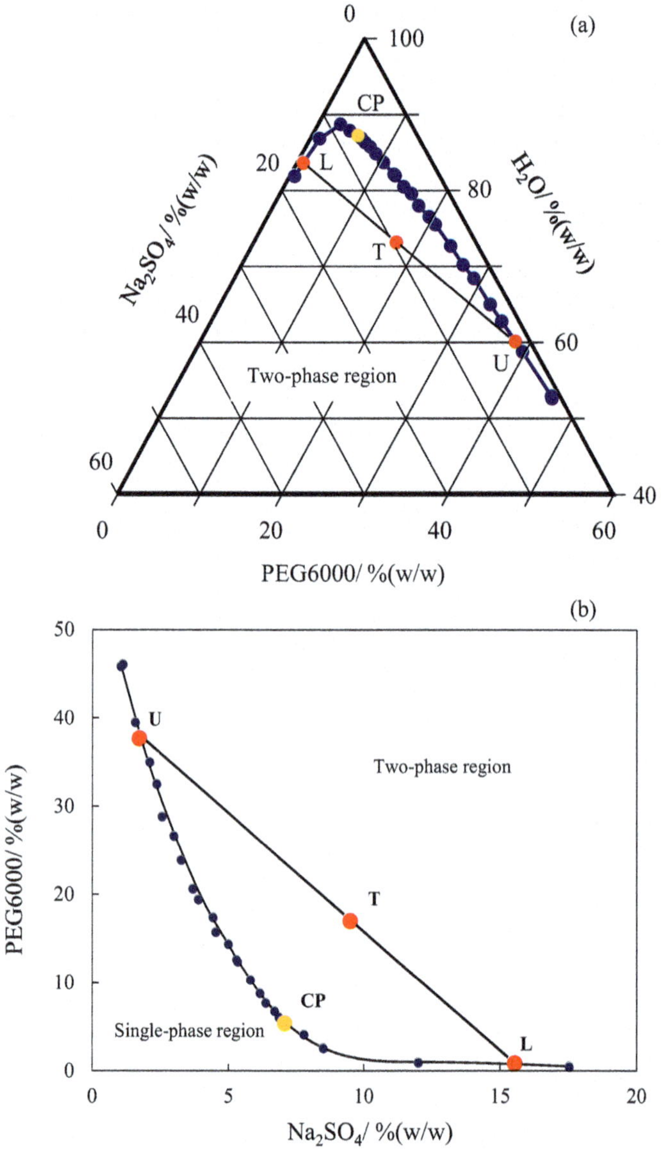

Fig. 1.1 Phase diagrams for the ternary system composed of PEG6000/Na_2SO_4/H_2O at 25 °C: **a** ternary representation and **b** orthogonal representation. CP represents the critical point of the diagram. T denotes the total composition of a random mixture in the two-phase region of the system, and U and B are the compositions of the upper and bottom phases of the system, respectively

phases, the compositions of which are represented by the points U (upper phase) and L (lower phase). In PEG/Na_2SO_4 ATPS, the less dense upper phase is the PEG-rich one, whereas the lower phase is the salt-rich one. This is not always the case and can vary with conditions such as temperature and salt concentration [25].

The straight line that connects the compositions of the total system (T) and those of upper and lower phases (U and L, respectively) is called tie-line and the composition difference between the two phases can be expressed by the tie-line length (TLL). TLL is calculated from the values of the polymer and salt weight fractions in the upper and lower phases as

$$\text{TLL} = \left(\Delta C_\text{p}^2 + \Delta C_\text{s}^2\right)^{1/2} \tag{1.1}$$

where ΔC_p and ΔC_s are the differences in polymer and salt compositions between the phases, respectively. TLL is thus an important parameter in the separation since it expresses the degree of divergence between the two phases. Mixtures with overall compositions along a specific tie-line U-T-L have the same compositions of the two coexisting phases, but different mass or volume ratios of the phases. In Fig. 1.1b, the weight ratio of the upper and lower phases is equal to the ratio of TLL between T and L and that between U and T. As the TLL decreases, the difference in composition between the two phases becomes less and finally the compositions of both phases become equal at the critical point CP, and the system becomes a single homogeneous phase.

The phase diagram of ATPS is constructed by preparing the two-phase system and determining the concentrations of two components, polymer and salt for polymer/salt ATPS, in the coexisting phases by a suitable analytical method such as size-exclusion chromatography. The diagram can also be prepared visually by turbidmetry in which dropwise additions of a salt stock solution of known concentration are made to a solution of one phase of known composition until detection of a change in turbidity, indicating the formation of two liquid phases. Or conversely, known amounts of water are added to a tube containing a heterogeneous system until a homogeneous system is obtained while the tube is shaken. The binodal line can be determined for each system using different initial weights of polymer and salt stock solutions.

1.3 Structures and Properties of the Aqueous Polyethylene Glycol Phases as Extraction Media in ATPS

As described above, the PEG-based ATPS has a lot of advantages and is the most widely used ATPS for liquid–liquid extraction of various compounds. Therefore, the elucidation of the solute partitioning mechanism at the molecular level in these systems is of vital importance for their successful application.

Rogers et al. examined the partition behavior of various neutral low-molecular-weight organic compounds in ATPSs composed of PEGs of different molecular weights and some different kinds of inorganic salts and showed that the partition coefficients of organic compounds are solely determined by the degree of phase divergence of the ATPS, ΔC_p, or TLL [12, 13]. They also measured the partition coefficients for a wide range of organic compounds in a PEG/dextran ATPS and compared the values of the partition coefficients with those in a PEG/salt ATPS [16]. The results are well correlated between the two ATPS, although the partition coefficients obtained in the PEG/dextran ATPS are quite smaller than those in PEG/salt ATPS.

On the other hand, differential scanning calorimetry (DSC) studies on aqueous PEG solutions suggest that the separation selectivity of the PEG-based ATPS can be attributed to the partitioning of solute compounds into the hydrated PEG polymer structure [26–29]. These results indicate that the structures and properties of aqueous PEG solutions are predominant determinants of the partitioning of solutes in PEG-based ATPS.

PEGs are polymers with a broad range of technical applications and especially play an important role in modern biomaterials. Thus, the structures of PEGs and their aqueous solutions have been extensively investigated by a lot of experimental techniques [30–46] and computer simulations [46–54]. PEG is an amphiphilic polymer containing both polar ether groups (–O–) and nonpolar ethylene groups ($–CH_2CH_2–$), the former of which can form hydrogen bonds, while the latter may cause hydrophobic interaction [33, 37, 38, 42, 43, 47, 50]. It has been shown that PEG forms a flexible coil-like structure in the form of internal hydrated helices covered on the outside with ethylene groups [32, 35, 37, 38, 47, 51]. The number of water molecules hydrogen bonded to the polar site is reported to be around 2 per each oxygen although the value varies depending on the experimental technique or the definition applied to computer simulations [29, 32, 37, 44, 45, 48, 51, 54, 55]. These findings suggest that the aqueous solution of PEG has water structure consisting of water molecules hydrogen bonded to the polar site of the polymer and the hydrophobic hydration water structure around its nonpolar site in addition to the bulk water. It is thus necessary to take into account the molecular-scale heterogeneous structure of the PEG aqueous solution to clarify the solute partition mechanism in PEG-based ATPS.

It has been reported in recent years that PEGs form self-organized assembles, clusters, or vesicles, in aqueous solutions due to hydrophobic interactions caused by nonpolar ethylene groups [38, 40, 41]. So far, the solvent properties of aqueous solutions of PEGs in ATPS have usually been studied using a linear solvation energy relationship (LSER) based on Abraham's generalized solvation equation [12, 13, 16, 56–58]. According to this procedure, any property of a solute in a medium is described in terms of linear dependence through LSER on the solvent parameters representing the independent modes of solute–solvent interaction such as dipolarity/polarizability, hydrogen bond ability to donate and to accept a proton [59]. Although this methodology is superior to the usual method of expressing the total interaction of a solute with its environment in terms of a single polarity parameter, it cannot be applied to PEG solutions since the PEG solutions are essentially heterogeneous [59].

The observed solvent property or Abraham's solvent descriptor is a concentration-weighted average of those for the microphases in a PEG solution and does not give information on the heterogeneous structure of the solution.

The objective of Chap. 2 is to describe a recent successful attempt made to estimate heterogeneous structures in aqueous PEG solutions using Co(II) thiocyanate complex formation reaction as a probe. It has been shown that Co(II) distributes into the PEG-rich phase predominantly as a blue tetrahedral complex, $Co(NCS)_4^{2-}$, in PEG/salt ATPS [60–62]. This complex is formed in organic solvents such as *N*,*N*′-dimethylformamide [63], but hardly in aqueous solutions [60]. These findings imply that the investigation of complexation reactions of Co(II) with thiocyanate ion in aqueous PEG solutions provides information on the structures of PEGs in aqueous solutions and their roles in partition of solute compounds in PEG-based ATPS. In this chapter, the spectrophotometric study of the equilibria of the complex formation reactions in aqueous solutions of PEG with varying ethylene oxide chain length and concentration will be described [64]. The heterogeneous water structures in PEG solutions estimated by calculating the volume fractions of the individual microphases in the solutions and the intrinsic formation constant of the tetrahedral cobalt thiocyanate complex in each phase will also be shown.

1.4 Effects of the Background Phase-Forming Electrolytes on the Partition of Ionic Solutes in PEG/Salt ATPS

The partition of solute compounds in polymer-based ATPS depends on numerous factors originating from the polymers and inorganic salts to be used for forming the two phases. As well as the variables concerning the polymers, the type and concentration of the salts are among the most important factors in polymer/salt ATPS [12–14, 65–80], the most conventionally used ATPS for the separation of inorganic compounds. It has been demonstrated that the ionic composition in ATPS exerts pronounced effects on the partition of solute compounds, particularly on that of ionic solutes.

However, the effects of the salts added upon the partition of solute compounds are rather complex because the phase diagrams or the concentrations of the polymers in the coexisting phases depend on the types and concentrations of the salts. It has been shown that increasing the ability of salts to salt out polymers enlarges the two-phase area of polymer/salt ATPS in the phase diagram and shifts the binodal line towards the origin. The ability of ions salting-out polymer usually follows the Hofmeister series and can be related to the Gibbs free energy of hydration [12–14, 65–70, 75, 77]. This means that the ionic composition in the ATPS influences the partition of solutes not only by its own effect, but also by the effect on the polymer composition of the two phases. Therefore, the inherent effect of ions on the solute partitioning in ATPSs should be evaluated under the conditions where the effect of the polymer concentrations can be regarded as being constant [75].

In Chap. 3, the effects of the background phase-forming counterions and co-ions on the distribution coefficients of inorganic cations and anions in PEG/salt (Na_2SO_4, K_2HPO_4, and Na_2HPO_4) ATPSs will be discussed on the basis of a stoichiometric model presented for the partition of ionic solutes in two-phase systems. It will be shown that the effect of the type of phase-forming salts on the partition of ionic solutes can be interpreted by the ion partition model.

1.5 Trace Analyses of Metal Ions in Environmental Samples Based on the Separation and Preconcentration by Extraction with ATPS

The exploitation of ATPS for the enrichment and separation of inorganic compounds has attracted much attention in recent years and it has been shown that polymer/salt ATPS can be efficient for the separation of various metal ions [9, 11, 60–62, 72, 73, 81–97]. Although a lot of different water-soluble polymers may be utilized, PEGs or block copolymers composed of units of poly(ethylene oxide) and poly(propylene oxide) are exclusively used in combination with inorganic salts such as Na_2SO_4 to form ATPS for the separation of metal ions.

Extractions of metal ions with polymer/salt ATPS have been classified into three categories [9, 72, 81]: (1) extraction using a water-soluble extractant which coordinates the metal ion; (2) extraction by addition of an inorganic anion which forms a metal complex that partitions into the polymer-rich phase; and (3) extraction of metal oxoanions such as TcO_4^- and $HCrO_4^-$ by the polymer-rich phase alone without any extractant. Among these three categories, the second type of metal extraction will find wide application to soft metal ions, which form complexes with halide or pseudohalide anions. For example, Spivakov and his co-workers [82] first demonstrated that Cu(II), Co(II), Zn(II), Fe(III), In(III), and Mo(V) are extracted into the PEG-rich phase in a PEG/$(NH_4)_2SO_4$ system containing 1 mol L^{-1} NH_4SCN and 1–2 mol L^{-1} H_2SO_4 as thiocyanate complexes. It has also been shown that iodide ion is useful to extract several metal ions such as Cd(II) and Hg(II) [9, 60, 72, 83–88]. In these ATPSs, metal ions are extracted as anionic complexes of the inorganic anions in a similar way to the extraction of metal halide complexes with ethers or ketones [9, 60, 72]. The third category of metal extraction is also based on the preferential partitioning of anionic metal species, in the form of metal oxoanions, to the polymer-rich phase. Thus, the mechanisms of partition of the finally formed metal species for the second and third categories are essentially the same. However, the optimization of the conditions for selective extraction of the target metal ion is easier for the extraction of the second category than that of the third one since the concentration of the extractant, i.e., halide and pseudohalide ion, can be easily manipulated. On the other hand, the extraction technique of the third category can be used for speciation of the element in a sample. For example, Cr(III) is favorably partitioned to the salt-rich phase of the

ATPS, whereas the Cr(VI) is preferentially concentrated in the polymer-rich phase as $HCrO_4^-$ [89, 93, 94].

The first type of extraction is very commonly used in conventional liquid–liquid extractions using organic solvents. As most of the organic extraction reagents are not soluble in water, they are used as dissolved in organic solvents. On the other hand, the extracting agent must be water soluble to some extent in ATPS. Thus, there is a restriction on the use of certain organic ligands as extracting agents [11]. However, aqueous solutions containing PEG at high concentration are able to dissolve some water-insoluble extractants such as 1-(2-pyridylazo)-2-naphtol [95, 96] and dimethylglyoxime [97] for extraction of metal ions into the PEG-rich phase from the salt-rich phase because the PEG-rich phase has a hydrophobic domain in their structures.

So far, most studies have concerned the distribution behavior of metal species in ATPS and the applications of this technique to the analysis of trace metal ions in complex environmental or industrial samples are still relatively limited. Chapter 4 focuses on the extraction methods belonging to the second category and discusses the extraction mechanism of metal ions. Furthermore, analytical methods for the determination of trace amounts of metal ions in environmental samples are presented.

1.6 Multistep pH-Peak-Focusing Countercurrent Chromatography

Separation with ATPS can be achieved by a chromatographic technique as well as a conventional batch procedure. Countercurrent chromatography (CCC) is a chromatographic method using partitioning of solutes between two immiscible liquid phases without any solid support. It is thus relatively simple to introduce ATPS into CCC as a separation platform. Among all the existing CCC apparatuses, the hydrodynamic CCC developed by Ito and his co-workers is the most advanced form and has now been accepted as an efficient preparative technique, which yields a highly efficient separation of a large amount of samples [98–100]. Although ATPS has been used in CCC mainly for the separation and purification of organic substances [101–105], CCC with ATPS can also be applied for separation of inorganic compounds [89, 106, 107].

pH-peak-focusing is a unique technique introduced in CCC from a finding that a thyroxine analog produced an unusually sharp elution peak [108, 109]. This technique is based on neutralization during elution between the mobile phase and the stationary phase of which the pH values are different from each other. The neutralization is initiated at the mobile phase front but advances through the column at a velocity lower than that of the mobile phase, forming a sharp pH jump border between basic and acidic zones in the mobile and stationary phases. In pH-peak-focusing CCC, acidic and basic compounds in a sample solution are concentrated during elution at the sharp pH border in the column [105–110].

On the other hand, Shimizu et al. have developed multistep pH-peak-focusing CCC that enables the separation of metal ions while maintaining on-column enrichment using PEG/Na_2SO_4 ATPS [106]. An important requirement for performing multistep pH-peak-focusing is that both of the stationary phase and the mobile phase should have large capacity for both of the acid and base forms of buffering agents. Since the main component of the ATPS is water, generally used hydrophilic buffering agents can distribute between the two phases even if they are ionic species. Therefore, ATPS can fulfill the requirement for performing multistep pH-peak-focusing CCC. Chapter 5 presents the mechanism of multistep pH-peak-focusing CCC and demonstrates an application of this technique to separation and enrichment of rare earth elements.

1.7 Aqueous Liquid Chromatography with Hydrophilic Polymer Gel Columns

Despite the advantages described above over traditional liquid–liquid extraction methods with organic solvents, ATPS has not found wider acceptance [11, 111]. Some of the probable reasons include the difficulty of separation of the target compounds from either the polymer or salt-rich phase after the extraction and considerable loss of phase-forming components (polymers and salts) since ATPS contains high concentrations of polymers and salts.

One of the approaches to overcome these problems is the use of an immobilized polymer phase [111–116]. Polymer chains anchored to the surface of a solid support or cross-linked hydrophilic polymer gels can sorb water and swell, but do not dissolve in solution. Thus, the immobilized polymer phases are predicted to function similarly to the polymer-rich phase of ATPS without additional manipulation for phase separation and reduce the problems related to the isolation of the target compound from the polymer matrix and the loss of the polymer components used to form the ATPS.

Rogers et al. developed a solid PEG-based ATPS resin by grafting high molecular weight PEGs to polystyrene-divinylbenzene copolymer particles and performed the ATPS extraction of pertechnetate ion and mercury(II) using the resin [111, 113]. They showed that pertechnetate ion and mercury(II) accumulated into the resin from salt solutions can be stripped off the resin by elution with water.

On the other hand, a lot of kinds of cross-linked hydrophilic polymer gels are known to exhibit different sorption properties for various inorganic and organic compounds and are widely used as separation membranes for dialysis and reverse osmosis and column packing materials for liquid chromatography [26, 27, 115, 117–127]. The mechanism of separation using these polymer gel materials have been intensely investigated and it has been revealed that separation selectivities exerted by a water-swollen polymer gel and the polymer-based ATPS are approximately the same; the separation can be attributed to partition of solute compounds into the hydrated polymer structure [26]. In Chap. 6, the solute separation selectivity shown by various water-swollen hydrophilic polymer gels used as packing materials for liquid chromatography columns will be discussed in relation to the state of water.

1.8 ATPS Chromatographic Techniques for Separation and Enrichment of Metal Ions and Inorganic Anions: Multistep pH-Peak-Focusing Liquid Chromatography and Partition/Ion-Exclusion Chromatographic Ion Stacking

Recently the author's research group presented two ATPS chromatographic techniques for separation and enrichment of metal ions and inorganic anions, multistep pH-peak-focusing liquid chromatography [128] and partition/ion-exclusion chromatographic ion stacking (PIEC ion stacking) [129, 130]. Multistep pH-peak-focusing liquid chromatography was developed using a hydrophilic polymer gel column, which shows dramatically improved separation efficiency compared with multistep pH-peak-focusing CCC. On the other hand, PIEC ion stacking is a novel ion enrichment method developed based on the distribution mechanism of ions in liquid chromatography with ATPS. The discovery of new principles sometimes gives rise to the development of new analytical methods. These two liquid chromatographic techniques have not yet reached a state of perfection and require much improvement. However, they have great potential as environmentally benign analytical methods that complement conventional methods. In the last chapter, the fundamental mechanisms and practical applications of multistep pH-peak-focusing liquid chromatography and PIEC ion stacking with hydrophilic polymer gel columns will be described as efficient separation and enrichment techniques based on ATPS.

References

1. Beijerinck MW (1896) Über eine eigentiimlichkeit der löslichen stärke. Zentbl Bakteriol Parasitenkd Infektkrankh 2:697–699
2. Albertsson PA (1986) Partitioning of cell particles and macromolecules, 3rd edn. Wiley-Interscience, New York
3. Albertsson PA (1958) Partition of proteins in liquid polymer-polymer two-phase systems. Nature 182:709–711
4. David E, Niculescu V-C (2021) Volatile organic compounds (VOCs) as environmental pollutants: occurrence and mitigation using nanomaterials. Int J Environ Res Publ Health 18:13147
5. Fruijtier-Pölloth C (2005) Safety assessment on polyethylene glycols (PEGs) and their derivatives as used in cosmetic products. Toxicol 214:1–38
6. Hermansky SJ, Leung HW (1997) Cutaneous toxicity studies with methoxy polyethylene glycol-350 (MPEG-350) in rats and rabbits. Food Chem Toxicol 35:1031–1039

7. Prentice DE, Majeed SK (1978) Oral toxicity of polyethylene glycol (PEG 200) in monkeys and rats. Toxicol Lett 2:119–122
8. Gullapalli RP, Mazzitelli CL (2015) Polyethylene glycols in oral and parenteral formulations—a critical review. Int J Pharm 496:219–239
9. Rogers RD, Eiteman MA (eds) (1995) Aqueous biphasic separations: biomolecules to metal ions. Plenum Press, New York
10. Saha N, Sarkar B, Sen K (2022) Aqueous biphasic systems: a robust platform for green extraction of biomolecules. J Mol Liq 363:119882
11. Karmakar R, Sen K (2019) Aqueous biphasic extraction of metal ions: an alternative technology for metal regeneration. J Mol Liq 273:231–247
12. Willauer HD, Huddleston JG, Rogers RD (2002) Solute partitioning in aqueous biphasic systems composed of polyethylene glycol and salt: the partitioning of small neutral organic species. Ind Eng Chem Res 41:1892–1904
13. Willauer HD, Huddleston JG, Rogers RD (2002) Solvent properties of aqueous biphasic systems composed of polyethylene glycol and salt characterized by the free energy of transfer of a methylene group between the phases and by a linear solvation energy relationship. Ind Eng Chem Res 41:2591–2601
14. da Silva LHM, da Silva MCH, de Aquino RAN, Francisco KR, Cardoso MVC, Minim LA, Coimbra JSR (2006) Nitroprusside-PEO enthalpic interaction as a driving force for partitioning of the $[Fe(CN)_5NO]^{2-}$ anion in aqueous two-phase systems formed by poly(ethylene oxide) and sulfate salts. J Phys Chem B 110:23540–23546
15. Chen J, Spear SK, Huddleston JG, Rogers RD (2005) Polyethylene glycol and solutions of polyethylene glycol as green reaction media. Green Chem 7:64–82
16. Moody ML, Willauer HD, Griffin ST, Huddleston JG, Rogers RD (2005) Solvent property characterization of poly(ethylene glycol)/dextran aqueous biphasic systems using the free energy of transfer of a methylene group and a linear solvation energy relationship. Ind Eng Chem Res 44:3749–3760
17. Freire MG (ed) (2016) Ionic-liquid-based aqueous biphasic systems: fundamentals and applications. Springer, Berlin
18. Ranke J, Stolte S, Störmann R, Arning J, Jastorff B (2007) Design of sustainable chemical products: the example of ionic liquids. Chem Rev 107:2183–2206
19. Meng J, Wang Y, Zhou Y, Chen J, Wei X, Ni R, Liu Z, Xu F (2019) Development of different deep eutectic solvent aqueous biphasic systems for the separation of proteins. RSC Adv 9:14116
20. Xu K, Wang Y, Huang Y, Li N, Wen Q (2015) A green deep eutectic solvent-based aqueous two-phase system for protein extracting. Anal Chim Acta 864:9–20
21. Zhang H, Wang Y, Zhou Y, Chen J, Wei X, Xu P (2018) Aqueous biphasic systems formed by deep eutectic solvent and new-type salts for the high-performance extraction of pigments. Talanta 181:210–216
22. Marchel M, Coroadinha AS, Marrucho IM (2020) Novel acidic deep eutectic solvent-based aqueous biphasic systems for efficient extraction of pepsin. ACS Sustain Chem Eng 8:12400–12408
23. Capela EV, Santos JHPM, Boal-Palheiros I, Coutinho JAP, Ventura SPM, Freire MG (2019) A simple approach for the determination and characterization of ternary phase diagrams of aqueous two-phase systems composed of water, polyethylene glycol and sodium carbonate. Chem Eng Educ 53:112–120
24. MacCarthy P (1983) Ternary and quaternary composition diagrams: an overview of the subject. J Chem Educ 60:922–928
25. Eiteman MA (1994) Temperature-dependent phase inversion and its effect on partitioning in the poly(ethylene glycol)-ammonium sulfate aqueous two-phase system. J Chromatogr A 668:13–19
26. Shibukawa M, Ichikawa R, Baba T, Sakamoto R, Saito S, Oguma K (2008) Separation selectivity of aqueous polyethylene glycol-based separation systems: DSC, LC and aqueous two-phase extraction studies. Polymer 49:4168–4173

27. Baba T, Sakamoto R, Shibukawa M, Oguma K (2004) Solute retention and the states of water in polyethylene glycol and poly(vinyl alcohol) gels. J Chromatogr A 1040:45–51
28. Liang Y, Xu S, Shang Y, Peng C, Liu H (2014) Aqueous four-phase system of SDS/DTAB/PEG/NaBr/H_2O. Colloids Surf 454:152–158
29. Berton P, Kelley SP, Bridges NJ, Klingshirn MA, Huddleston JG, Willauer HD, Baldwin JW, Moody ML, Rogers RD (2019) Water in solutions of chaotropic and kosmotropic salts: a differential scanning calorimetry investigation. J Chem Eng Data 64:4781–4792
30. Begum R, Matsuura H (1997) Conformational properties of short poly(oxyethylene) chains in water studied by IR spectroscopy. J Chem Soc Faraday Trans 93:3839–3848
31. Rozenberg M, Loewenschuss A, Marcus Y (1998) IR spectra and hydration of short-chain polyethyleneglycols. Spectrochim Acta Part A 54:1819–1826
32. Oesterhelt F, Rief M, Gaub HE (1999) Single molecule force spectroscopy by AFM indicates helical structure of poly(ethylene-glycol) in water. New J Phys 1:6
33. Hammouda B, Ho D, Kline S (2002) SANS from poly(ethylene oxide)/water systems. Macromolecules 35:8578–8585
34. Chen C, Even MA, Wang J, Chen Z (2002) Sum frequency generation vibrational spectroscopy studies on molecular conformation of liquid polymers poly(ethylene glycol) and poly(propylene glycol) at different interfaces. Macromolecules 35:9130–9135
35. Hammouda B, Ho DL, Kline S (2004) Insight into clustering in poly(ethylene oxide) solutions. Macromolecules 37:6932–6937
36. Pochylski M, Aliotta F, Blaszczak Z, Gapiński J (2006) Structuring effects and hydration phenomena in poly(ethylene glycol)/water mixtures investigated by Brillouin scattering. J Phys Chem B 110:20533–20539
37. Shephard JJ, Bremer PJ, McQuillan AJ (2009) Structure and conformation in mixtures of methyl-terminated poly(ethylene oxide) and water. Principal component analysis and band fitting of infrared absorptions. J Phys Chem B 113:14229–14238
38. Azri A, Giamarchi P, Grohens Y, Olier R, Privat M (2012) Polyethylene glycol aggregates in water formed through hydrophobic helical structures. J Colloid Interf Sci 379:14–19
39. Cao N, Zhao Y, Chen H, Huang J, Yu M, Bao Y, Wang D, Cui S (2022) Poly(ethylene glycol) becomes a supra-polyelectrolyte by capturing hydronium ions in water. Macromolecules 55:4656–4664
40. Titus AR, Madeira PP, Ferreira LA, Chernyak VY, Uversky VN, Zaslavsky BY (2022) Mechanism of phase separation in aqueous two-phase systems. Int J Mol Sci 23:14366
41. Paik SP, Ghatak SK, Dey D, Sen K (2012) Poly(ethylene glycol) vesicles: self-assembled site for luminescence generation. Anal Chem 84:7555–7561
42. Sailaja D, Raju KN, Santha devi GS, Subbarangaiah K, (1998) Ultrasonic behaviour of aqueous solutions of polyethylene glycols on the temperature of adiabatic compressibility minimum of water. Eur Polym J 34:887–890
43. Marinov VS, Matuura H (2002) Raman spectroscopic study of temperature dependence of water structure in aqueous solutions of poly(oxyethylene) surfactant. J Mol Struct 610:105–112
44. Tominaga T, Hishida M, Murakami D, Fujii Y, Tanaka M, Seto H (2022) Experimental evidence of slow mode water in the vicinity of poly(ethylene oxide) at physiological temperature. J Phys Chem B 126:1758–1767
45. Shikata T, Okuzono M, Sugimoto N (2013) Temperature-dependent hydration/dehydration behavior of poly(ethylene oxide)s in aqueous solution. Macromolecules 46:1956–1961
46. Rubinson KA, Krueger S (2009) Poly(ethylene glycol)s 2000–8000 in water may be planar: a small-angle neutron scattering (SANS) structure study. Polymer 50:4852–4858
47. Borodin O, Bedrov D, Smith GD (2002) Molecular dynamics simulation study of dielectric relaxation in aqueous poly(ethylene oxide) solutions. Macromolecules 35:2410–2412
48. Tasaki K (1996) Poly(oxyethylene)-water interactions: a molecular dynamics study. J Am Chem Soc 118:8459–8469
49. Smith GD, Bedrov D (2002) A molecular dynamics simulation study of the influence of hydrogen-bonding and polar interactions on hydration and conformations of a poly(ethylene oxide) oligomer in dilute aqueous solution. Macromolecules 35:5712–5719

50. Xu Q, Mi J, Zhong C (2010) Structure of poly(ethylene glycol)-water mixture studied by polymer reference interaction site model theory. J Chem Phys 133:174104
51. Dahal UR, Dormidontova EE (2017) The dynamics of solvation dictates the conformation of polyethylene oxide in aqueous, isobutyric acid and binary solutions. Phys Chem Chem Phys 19:9823–9832
52. Quezada GR, Toro N, Saavedra J, Robles P, Salazar I, Navarra A, Jeldres RI (2021) Molecular dynamics study of the conformation, ion adsorption, diffusion, and water structure of soluble polymers in saline solutions. Polymers 13:3550
53. Smith GD, Borodin O, Bedrov D (2002) A revised quantum chemistry-based potential for poly(ethylene oxide) and its oligomers in aqueous solution. J Comput Chem 23:1480–1488
54. Dormidontova EE (2002) Role of competitive PEO-water and water-water hydrogen bonding in aqueous solution PEO behavior. Macromolecules 35:987–1001
55. Smith GD, Bedrov D, Borodin O (2000) Molecular dynamics simulation study of hydrogen bonding in aqueous poly(ethylene oxide) solutions. Phys Rev Lett 85:5583
56. Huddleston JG, Willauer HD, Rogers RD (2000) Solvatochromic studies in polyethylene glycol-salt aqueous biphasic systems. J Chromatogr B 743:137–149
57. da Silva NR, Ferreira LA, Madeira PP, Teixeira JA, Uversky VN, Zaslavsky BY (2015) Analysis of partitioning of organic compounds and proteins in aqueous polyethylene glycol-sodium sulfate aqueous two-phase systems in terms of solute-solvent interactions. J Chromatogr A 1415:1–10
58. Titus AR, Ferreira LA, Belgovskiy AI, Kooijman EE, Mann EK, Mann JA Jr, Meyer WV, Smart AE, Uversky VN, Zaslavsky BY (2020) Interfacial tension and mechanism of liquid-liquid phase separation in aqueous media. Phys Chem Chem Phys 22:4574–4580
59. Bagchi S (2015) Solvation in heterogeneous media. In: Birdi KS (ed) Handbook of surface and colloid chemistry, 4th edn. CRC Press, London, pp 461–480
60. Shibukawa M, Nakayama N, Hayashi T, Shibuya D, Endo Y, Kawamura S (2001) Extraction behaviour of metal ions in aqueous polyethylene glycol-sodium sulphate two-phase systems in the presence of iodide and thiocyanate ions. Anal Chim Acta 427:293–300
61. da Patricio PR, Mesquita MC, da Silva LHM, da Silva MCH (2011) Application of aqueous two-phase systems for the development of a new method of cobalt(II), iron(III) and nickel(II) extraction: a green chemistry approach. J Hazard Mater 193:311–318
62. Shirayama S, Uda T (2015) Recovery of cobalt ion into polyethyleneglycol (PEG) gel phase as thiocyanato complex. Mater Trans 56:610–616
63. Ishiguro S, Ozutsumi K (1990) Thermodynamics and structure of isothiocyanato complexes of manganese(II), cobalt(II), and nickel(II) ions in N,N-dimethylformamide. Inorg Chem 29:1117–1123
64. Shibukawa M, Hirasawa T, Saito S (2024) Structure of polyethylene glycol/water mixtures as a separation medium of aqueous two-phase extraction systems studied by equilibrium analysis of thiocyanato complexation of cobalt(II). Polymer 293:126668
65. Bailey Jr FE, Callard RW (1959) Some properties of poly(ethylene oxide) in aqueous solution. J Appl Polym Sci 1:56–62
66. Boucher EA, Hines PM (1976) Effects of inorganic salts on the properties of aqueous poly(ethylene oxide) solutions. J Polym Sci Polym Phys Ed 14:2241–2251
67. Ataman M, Boucher EA (1982) Properties of aqueous salt solutions of poly(ethylene oxide). J Polym Sci Polym Phys Ed 20:1585–1592
68. Florin E, Kjellander R, Eriksson JC (1984) Salt effects on the cloud point of the poly(ethylene oxide) + water system. J Chem Soc, Faraday Trans 1 80:2889–2910
69. Ananthapadmanabhan KP, Goddard ED (1986) A correlation between clouding and aqueous biphase formation in polyethylene oxide/inorganic salt systems. J Colloid Interface Sci 113:294–296
70. Ananthapadmanabhan KP, Goddard ED (1987) Aqueous biphase formation in polyethylene oxide-inorganic salt systems. Langmuir 3:25–31
71. Berggren K, Johansson HO, Tjerneld F (1995) Effects of salts and the surface hydrophobicity of proteins on partitioning in aqueous two-phase systems containing thermoseparating ethylene oxide-propylene oxide copolymers. J Chromatogr A 718:67–79

72. Rogers RD, Bond AH, Bauer CB, Zhang J, Griffin ST (1996) Metal ion separations in polyethylene glycol-based aqueous biphasic systems: correlation of partitioning behavior with available thermodynamic hydration data. J Chromatogr B 680:221–229
73. Rogers RD, Zhang J (1996) Effects of increasing polymer hydrophobicity on distribution ratios of TcO_4^- in polyethylene/poly(propylene glycol)-based aqueous biphasic systems. J Chromatogr B 680:231–236
74. da Silva LHM, Loh W (2000) Calorimetric investigation of the formation of aqueous two-phase systems in ternary mixtures of water, poly(ethylene oxide) and electrolytes (or dextran). J Phys Chem 104:10069–10073
75. Shibukawa M, Matsuura K, Shinozuka Y, Mizuno S, Oguma K (2000) Effects of phase-forming cations and anions on the partition of ionic solutes in aqueous polyethylene glycol-inorganic salt two-phase systems. Anal Sci 16:1039–1044
76. Ferreira LA, Teixeira JA (2011) Salt effect on the (polyethylene glycol 8000 + sodium sulfate) aqueous two-phase system: relative hydrophobicity of the equilibrium phases. J Chem Thermodyn 43:1299–1304
77. Ferreira LA, Teixeira JA (2011) Salt effect on the aqueous two-phase system PEG 8000-sodium sulfate. J Chem Eng Data 56:133–137
78. Silvério SC, Wegrzyn A, Lladosa E, Rodríguez O, Macedo EA (2012) Effect of aqueous two-phase system constituents in different poly(ethylene glycol)-salt phase diagrams. J Chem Eng Data 57:1203–1208
79. Hespanhol MC, da Silva LHM, Fontoura BM, de Andrade VM, Mageste AB, de Lemos LR (2019) Phase diagrams, densities, and refractive indexes of aqueous two-phase systems comprising (F68, L64, or PEO1500) + (ammonium, sodium, or potassium thiocyanate salts) + water: effect of cation and type of macromolecule. J Chem Eng Data 64:1991–1998
80. Milevskiy NA, Boryagina IV, Karpukhina EA, Kuznetsov VN, Kabanova EG (2021) Effect of sodium chloride and pH on the composition of the equilibrium phases and the partition of palladium(II) in the aqueous two-phase system PEG1500-Na_2SO_4-water. J Chem Eng Data 66:1021–1031
81. Rogers RD, Bond AH, Bauer CB (1993) Metal ion separations in polyethylene glycol-based aqueous biphasic systems. Sep Sci Technol 28:1091–1126
82. Zvarova TI, Shkinev VM, Vorob'eva GA, Spivakov BY, Zolotov YA (1984) Liquid-liquid extraction in the absence of usual organic solvents: application of two-phase aqueous systems based on a water-soluble polymer. Microchim Acta 84:449–458
83. Bulgariu L, Bulgariu D (2008) Extraction of metal ions in aqueous polyethylene glycol–inorganic salt two-phase systems in the presence of inorganic extractants: correlation between extraction behaviour and stability constants of extracted species. J Chromatogr A 1196–1197:117–124
84. Hamta A, Dehghani MR (2017) Application of polyethylene glycol based aqueous two-phase systems for extraction of heavy metals. J Mol Liq 231:20–24
85. Bulgariu L, Bulgariu D (2006) Hg (II) extraction in a PEG-based aqueous two-phase system in the presence of halide ions. I. Liquid phase analysis. Central Eur J Chem 4:246–257
86. Bulgariu L, Bulgariu D, Sârghie I, Măluţan T (2007) Cd(II) Extraction in PEG-based two-phase aqueous systems in the presence of iodide ions. Analysis of PEG-rich solid phases. Central Eur J Chem 5:291–302
87. Bulgariu L, Bulgariu D (2013) Selective extraction of Hg(II), Cd(II) and Zn(II) ions from aqueous media by a green chemistry procedure using aqueous two-phase systems. Sep Purif Technol 118:209–216
88. Shibukawa M, Nobushima D, Sakuma S, Sasaki M, Nakamura K, Matsutani Y, Saitoh K, Saito S (2016) Selective spectrophotometric determination of trace amounts of cadmium in soil and sediment samples using a green aqueous biphasic extraction. Anal Sci 32:1095–1100
89. Ishii K, Tanaka Y, Hata K, Goto M, Saitoh K, Minamisawa H, Shibukawa M, (2004) Separation of inorganic ions by high-speed countercurrent chromatography with aqueous biphasic system. Bunseki Kagaku 53:911–917

90. Sun P, Huang K, Lin J, Liu H (2018) Role of hydrophobic interaction in driving the partitioning of metal ions in a PEG-based aqueous two-phase system. Ind Eng Chem Res 57:11390–11398
91. Patrício PR, Quintão JC, da Silva LHM, Hespanhol MC (2022) One-pot separation and preconcentration for the determination of chromium species using an aqueous biphasic system. J Braz Chem Soc 33:978–986
92. Spear SK, Griffin ST, Huddleston JG, Rogers RD (2000) Radiopharmaceutical and hydrometallurgical separations of perrhenate using aqueous biphasic systems and the analogous aqueous biphasic extraction chromatographic resins. Ind Eng Chem Res 39:3173–3180
93. Patrício PR, Cunha RC, Vargas SJR, Coelho YL, da Silva LHM, da Silva MCH (2016) Chromium speciation using aqueous biphasic systems: development and mechanistic aspects. Sep Purif Technol 158:144–154
94. Roy K, Lahiri S (2008) Species dependent radiotracer study of Cr(VI) and Cr(III) using an aqueous biphasic system. Radiochim Acta 96:49–53
95. Leite DS, Carvalho PLG, de Lemos LR, Mageste AB, Rodrigues GD (2017) Hydrometallurgical separation of copper and cobalt from lithium-ion batteries using aqueous two-phase systems. Hydrometallurgy 169:245–252
96. de Lemos LR, Santos IJB, Rodrigues GD, da Silva LHM, da Silva MCH (2012) Copper recovery from ore by liquid–liquid extraction using aqueous two-phase system. J Hazard Mater 237–238:209–214
97. Yoshikuni N, Baba T, Tsunoda N, Oguma K (2005) Aqueous two-phase extraction of nickel dimethylglyoximato complex and its application to spectrophotometric determination of nickel in stainless steel. Talanta 66:40–44
98. Ito Y, Conway WD (eds) (1996) High-speed countercurrent chromatography. John Wiley & Sons, New York
99. Berthod A, Maryutina T, Spivakov B, Shpigun O, Sutherland IA (2009) Countercurrent chromatography in analytical chemistry. Pure Appl Chem 81:355–387
100. Ito Y (2005) Countercurrent chromatography. In: Worsfold P, Townshend A, Poole C (eds) Encyclopedia of analytical science, 2nd edn. Elsevier, Amsterdam, pp 241–252
101. Shinomiya K, Kabasawa Y, Yanagidaira K, Sasaki H, Muto M, Okada T, Ito Y (2003) Protein separation by nonsynchronous coil planet centrifuge with aqueous–aqueous polymer phase systems. J Chromatogr A 1005:103–112
102. Shibusawa Y, Kinebuchi T, Shimizu M, Shindo H, Ito Y (2003) Purification of single-strand DNA binding protein from an *Escherichia coli* lysate using counter-current chromatography, partition and precipitation. J Chromatogr B 793:275–279
103. Yanagida A, Isozaki M, Shibusawa Y, Shindo H, Ito Y (2004) Purification of glucosyltransferase from cell-lysate of Streptococcus mutans by counter-current chromatography using aqueous polymer two-phase system. J Chromatogr B 805:155–160
104. Ding J, Li S, Zhao Y, Guan YH, Deng L, Deng Q (2017) Properties of hydrodynamic J-type countercurrent chromatography for protein separation using aqueous two-phase systems: with special reference to constructing conical columns. J Chromatogr A 1499:103–110
105. Song G, Du Q (2010) Isolation of a polysaccharide with anticancer activity from Auricularia polytricha using high-speed countercurrent chromatography with an aqueous two-phase system. J Chromatogr A 1217:5930–5934
106. Shimizu K, Kuribayashi H, Watanabe H, Shimasaki T, Azuma K, Horie Y, Saitoh K, Saito S, Shibukawa M (2013) Multistep pH-peak-focusing countercurrent chromatography with a polyethylene glycol-Na_2SO_4 aqueous two phase system for separation and enrichment of rare earth elements. Anal Chem 85:978–984
107. Mokhodoeva O, Rudik I, Shkinev V, Maryutina T (2021) Countercurrent chromatography approach to palladium and platinum separation using aqueous biphasic system. J Chromatogr A 1657:462581
108. Ito Y, Shibusawa Y, Fales HM, Cahnmann HJ (1992) Studies on an abnormally sharpened elution peak observed in counter-current chromatography. J Chromatogr 625:177–181
109. Ito Y, Oka H (2009) pH-peak-focusing and pH-zone-refining CCC. In: Cazes J (ed) Encyclopedia of chromatography, 3rd edn. CRC Press, London, pp 1080–1813

110. Wang C, Zuo G, Wang X, Kim HY, Zhao S, Sun W, Tong S, Lim SS (2020) Retention mechanism of pH-peak-focusing in countercurrent chromatographic separation of baicalin and wogonoside from *Scutellaria baicalensis* Georgi. J Sep Sci 43:3806–3815
111. Dilip M, Griffin ST, Spear SK, Rijksen C, Rodríguez H, Rogers RD (2008) Dual nature of polyethylene glycol-based aqueous biphasic extraction chromatographic (ABEC) resins: uptakes of perchlorate versus mercury(II). Ind Eng Chem Res 47:7390–7396
112. Huddleston JG, Willauer HD, Boaz K, Rogers RD (1998) Separation and recovery of food coloring dyes using ABEC resins. J Chromatogr B 11:237–244
113. Rogers RD, Griffin ST (1998) Partitioning of mercury in aqueous biphasic systems and on ABEC^{TM} resins. J Chromatogr B 711:277–283
114. Griffin ST, Dilip M, Spear SK, Huddleston JG, Rogers RD (2006) The opposite effect of temperature on polyethylene glycol-based aqueous biphasic systems versus aqueous biphasic extraction chromatographic resins. J Chromatogr B 844:23–31
115. Huddleston JG, Looney TK, Broker GA, Griffin ST, Spear SK, Rogers RD (2003) Comparative behavior of poly(ethylene glycol) hydrogels and poly(ethylene glycol) aqueous biphasic systems. Ind Eng Chem Res 42:6088–6095
116. Mo YL, Fischlschweiger M, Zeiner T (2022) Polymeric aqueous two phase system with an immobilized phase by cross-linking. Ind Eng Chem Res 61:15381–15389
117. Shibukawa M, Ohta N (1988) Exclusion chromatography of inorganic compounds. In: Dubin PL (ed) Aqueous size-exclusion chromatography. Elsevier, Amsterdam, pp 77–115
118. Hoffman AS (2002) Hydrogels for biomedical applications. Adv Drug Delivery Rev 43:3–12
119. Sassi AP, Shaw AJ, Han SM, Blanch HW, Prausnitz JM (1996) Partitioning of proteins and small biomolecules in temperature and pH sensitive hydrogels. Polymer 37:2151–2164
120. Shibukawa M, Ohta N, Kuroda R (1981) Distribution processes of inorganic solutes in gel chromatography. Anal Chem 53:1620–1627
121. Shibukawa M, Ohta N (1986) Model for partition chromatography of ionic solutes in the presence of eluent electrolytes. Chromatographia 22:261–267
122. Shibukawa M, Ohta N (1988) A new method for the determination of mobile phase volume in normal and reversed-phase liquid chromatography. Chromatographia 25:288–294
123. Shibukawa M, Ohta N, Onda N (1990) Investigation of the states of water in water-swollen hydrogels by liquid chromatography and differential scanning calorimetry. Bull Chem Soc Jpn 63:3490–3494
124. Higuchi A, Iijima T (1985) D.s.c. investigation of the states of water in poly(vinyl alcohol-co-itaconic acid) membranes. Polymer 26:1833–1837
125. Hatakeyama T, Quinn FX (1994) Thermal analysis: fundamentals and applications to polymer science. Wiley, Chichester
126. Shibukawa M, Aoyagi K, Sakamoto R, Oguma K (1999) Liquid chromatography and differential scanning calorimetry studies on the states of water in hydrophilic polymer gel packings in relation to retention selectivity. J Chromatogr A 832:17–27
127. Shibukawa M (2006) Studies on the states of water in separation media of aqueous liquid chromatography in relation to separation selectivity. Bunseki Kagaku 55:149–162
128. Shibukawa M, Onoyama Y, Handa-Tasaki Y, Saito S (2024) Multistep pH-peak-focusing liquid chromatography with a hydrophilic polymer gel column for separation of rare earth elements. J Chromatogr A 1721:464829
129. Akter F, Ogiyama Y, Saito S, Shibukawa M (2017) Mechanism of ion stacking in aqueous partition chromatographic processes. J Sep Sci 40:3205–3213
130. Akter F, Saito S, Handa-Tasaki Y, Shibukawa M (2018) Partition/ion-exclusion chromatographic ion stacking for analysis of trace anions in water and salt samples by ion chromatography. Anal Sci 34:369–373

Chapter 2
Structures and Properties of the Aqueous Polyethylene Glycol Phases as Extraction Media in ATPS

Abstract PEG-based ATPSs have been successfully used for the separation of not only biomolecules but also inorganic compounds as an environmentally friendly platform for liquid–liquid extraction. However, a fundamental understanding of the structure of aqueous PEG solutions and its effect on the partitioning of solute compounds in the ATPS remains elusive. In this chapter, spectrophotometric studies on the formation equilibria of the complexes of Co(II) with thiocyanate ion in aqueous PEG solutions are described. From the dependence of the apparent formation constants of the tetrahedral and octahedral Co(II) thiocyanate complexes on the concentration of ethylene oxide (EO) and terminal hydroxyl (OH) groups in the solutions, a micro-heterogeneous structure consisting of pseudo-phases formed by hydration of the EO chains and OH groups in PEG molecules is estimated. It is revealed that the intrinsic formation constant of tetrahedral $Co(NCS)_4^{2-}$ complex in the EO pseudo-phase is much larger than those in the OH phase and the bulk water phase. An important conclusion is drawn from these studies that weakly hydrated anions such as $Co(NCS)_4^{2-}$ distribute into the hydrophobic outer region in the helices of the EO chains and the partitioning of solute molecules and ions in PEG-based ATPSs is governed by the difference in the fractions of the EO and OH phases between the upper and lower phases of the systems.

Keywords Aqueous two-phase system · Hydration structure of polyethylene glycol · Liquid–liquid extraction · Thiocyanato complexation of cobalt(II) · Micro-heterogeneous structure

2.1 Introduction

Polyethylene glycol (PEG) is the most commonly used phase-forming component for ATPS and the separation or preconcentration of various kinds of organic and inorganic compounds with PEG/salt and PEG/dextran ATPSs has been extensively investigated by many researchers [1–4]. However, we have not yet reached the stage where we can say that the separation mechanism has been clarified and the optimum

M. Shibukawa, *Polymer-Based Aqueous-Two Phase Systems*,
https://doi.org/10.1007/978-981-96-2888-9_2

separation conditions can be set based on the mechanism. Understanding the solute partitioning mechanism in PEG-based ATPS at the molecular level is thus critical for its successful application.

Rogers et al. investigated the partitioning of a series of nonionic substituted benzene compounds and aliphatic compounds in ATPSs composed of different molecular weights of PEG and different types of inorganic salts and showed that the distribution of organic solutes is governed only by the degree of phase divergence of the ATPS as expressed by the difference in PEG concentration between the two phases [5]. They also evaluated the relative hydrophobicities of several PEG/salt ATPSs by the free energy of transfer of a methylene group, ΔG_{CH_2}, measured utilizing a series of aliphatic alcohols and demonstrated that ΔG_{CH_2} is a function of the difference in PEG composition between the two phases represented in terms of PEG concentration or ethylene oxide concentration ($\Delta EO_{,m}$ represented in molarity scale) regardless of the type and concentration of the salt, or PEG molecular weight used to form the ATPS (Fig. 2.1) [6]. On the other hand, we compared the distribution behavior of *n*-alcohols, ketones, and nitroalkanes in aqueous liquid chromatography with a column packed with a PEG gel with that in ATPS formed from PEG and Na_2SO_4 or $(NH_4)_2SO_4$ and revealed that the separation selectivity exhibited by PEG gel systems and PEG-based ATPS can be regarded as similar [7]. These observations indicate that the structures and properties of the aqueous PEG-rich phase play a decisive role in the partitioning of solutes in PEG-based ATPS.

Linear solvation energy relationships (LSERs) have frequently been used to describe the intermolecular forces in the form of solute–solvent interactions involved

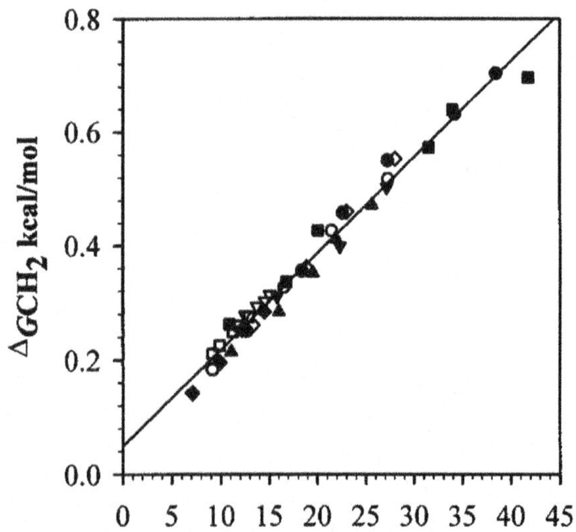

Fig. 2.1 ΔG_{CH_2} as a function of $\Delta EO_{,m}$ for a series of ATPS formed with 40% (w/w) PEG-2000 and increasing concentrations of salt [(●) K_3PO_4, (■) K_2CO_3, (▲) $(NH_4)_2SO_4$, (▼) NaOH, (◆ ◆) Li_2SO_4, (□) $MnSO_4$, (▽) $ZnSO_4$] or with K_3PO_4 and (○) 40% (w/w) PEG-1000 or (◇) 40% (w/w) PEG-3400 (reproduced from Ref. [6] by permission of The American Chemical Society)

in solute partitioning in ATPS [5, 6, 8–11] based on Abraham's generalized solvation equation [12–14]. According to this procedure, the change in the physicochemical parameter of a solute (S) due to solvation, $\Delta(S, i)$, is expressed in terms of a linear relationship with the solvent parameters representing the independent modes of solute–solvent interaction as shown in Eq. (2.1) [14]

$$\Delta(S, i) = P_0(S) + \sum a_\alpha(S)P_\alpha(i) \tag{2.1}$$

where i indicates a solvent, P_0 and a_α terms are the properties of the solute (S), and P_α (i) terms are the solvent properties pertinent to the α mode of interaction of the solute with a solvent.

Zaslavsky et al. [11] investigated the partition coefficient of a solute, K, in PEG-based ATPS using the following equation:

$$\log K = S_s\Delta\pi^* + B_s\Delta\alpha + A_s\Delta\beta + C_s c \tag{2.2}$$

where $\Delta\pi^*$, $\Delta\alpha$, $\Delta\beta$, and c are the differences between the solvent properties of the upper and lower phases in the ATPS, solvent dipolarity/polarizability, hydrogen-bond donor acidity, hydrogen-bond acceptor basicity, and electrostatic interactions, respectively. S_s, B_s, A_s, and C_s are the solute explanatory variables (the solute descriptors) that describe the contributions of the complementary interactions of the solute with the media in the coexisting two phases. They examined the partition coefficients of some ionic and nonionic organic compounds in PEG/dextran and PEG/Na_2SO_4 ATPSs formed with or without the addition of NaCl and $NaClO_4$ and investigated the relative contributions of different types of interactions using Eq. (2.2). It was shown from the result that hydrophobic and electrostatic properties and hydrogen bonding donor acidity are the most significant solvent properties of the systems. Rogers et al. [6] also developed a LSER relationship based on Abraham's equation using the partition coefficients measured for organic compounds of different structure and functional groups in PEG/salt ATPSs and concluded that the intermolecular solute–solvent interactions governing solute partitioning in a PEG/salt ATPS can be interpreted with the difference in hydrophobicity and hydrogen-bond donor acidity of the two phases.

The approaches based on LSER have considerable utility in aiding the prediction of solute partitioning in liquid–liquid extraction systems. However, it has been clarified recently that PEG aqueous solutions are heterogeneous due to the formation of self-organized aggregates, clusters, or vesicles of PEG molecules in aqueous solutions [15–17]. Therefore, the observed solvent properties are concentration or volume-weighted averages of the properties for the microphases in the two phases [14] and do not give information on their heterogeneous structure and the physicochemical property of each microphase. The measured value of P in Eq. (2.1) for a heterogeneous media can be assumed to be expressed by an average of the properties of the bulk aqueous phase (P_{bulk}) and microphases (P_i), as given by Eq. (2.3):

$$P = \frac{n_{\text{bulk}} P_{\text{bulk}} + \sum_i n_i P_i}{n_{\text{bulk}} + \sum_i n_i} \quad (2.3)$$

where n refers to moles of the solute, subscripts bulk and i denote the bulk aqueous phase and each microphase, respectively. The calculation of the value of P_{bulk} and P_i requires knowledge of n_{bulk} and n_i or the partition coefficient of the solute between the bulk phase and each microphase. This means that we can elucidate the structure of the heterogeneous solution phase by obtaining the quantitative data for the partitioning of a probe compound which can distinguish the microphase formed in the solution from the bulk aqueous phase.

It is known that the complexation behavior of Co(II) with thiocyanate ion strongly depends on the type of solvent and composition of the solution. Ishiguro and Ozutsumi demonstrated that the solution of Co(II) in *N,N′*-dimethylformamide (DMF) changes its color from pink to blue by adding thiocyanate ion even though this is rarely observed in aqueous solutions [18]. Fronæous and Larsson investigated the reaction of Co(II) and thiocyanate ion using water-methyl ethyl ketone (MEK) mixtures containing NaSCN and $Co(ClO_4)_2$ in mole proportions 5:1 and observed that the color of the solutions turns from red-pink to bright blue when the volume percent of MEK exceeds 20% [19]. These color changes of the Co(II) solutions have been clarified to be attributed to the change in coordination geometry from an octahedral aqua complex to tetrahedral $Co(NCS)_4^{2-}$ in aprotic solvent environments.

On the other hand, it has been reported that Co(II) is extracted as a blue complex into the PEG-rich phase in PEG/salt ATPSs by addition of thiocyanate ion [20–24]. The Co(II) complex extracted in a PEG/Na_2SO_4 ATPS was shown to be $Co(NCS)_4^{2-}$ from the investigation of the dependence of the distribution coefficient of Co(II) on the concentration of thiocyanate ion added to the ATPS [20]. Ishiguro et al. investigated the formation reactions of Co(II)-thiocyanato complexes in aqueous solutions of Triton X by spectrophotometry and calorimetry and showed that the octahedral $Co(NCS)(H_2O)_5^+$ and $Co(NCS)_2(H_2O)_4$ are formed in an aqueous phase, while $Co(NCS)_4^{2-}$ is formed in micelles of Triton X [25–28]. Triton X is a nonionic surfactant consisting of PEG and *tert*-octylphenylether and forms micelles having central hydrophobic cores made up of hydrocarbon chains. They also found that the formation of $Co(NCS)_4^{2-}$ in micelles is enhanced with increasing PEG molecular weight and concluded that the tetrahedral cobalt complex is formed at ethylene oxide moieties as well as octylphenyl micelle cores [28].

These findings that Co(II) forms the tetrahedral complex with thiocyanate ion in PEG-rich phases in PEG/salt ATPSs as well as in micelles of Triton X and organic solvents such as DMF [18] suggest that the complexation reactions of Co(II) with thiocyanate ion in aqueous PEG solutions can be a probe giving the information on the structures of aqueous PEG solutions and their roles in solute distribution in PEG-based ATPS.

In this chapter, the structure of the aqueous PEG solution revealed by the formation reaction of Co(II) thiocyanate complexes will be described [29]. The equilibria of the complexation reactions were investigated by spectrophotometry with varying the molecular weight and concentration of PEG in aqueous solutions. The heterogeneous microstructure of an aqueous PEG solution is estimated by calculating the volume fractions of the individual microphases formed at the moieties of PEG molecules and the intrinsic formation constant of the tetrahedral cobalt thiocyanate complex in each phase. The distribution of solute molecules and ions in the heterogeneous structure of PEG solution is discussed in comparison with those estimated by differential scanning calorimetry (DSC), dynamic light scattering (DLS), and molecular dynamics simulation studies.

2.2 Complexation Reactions of Co(II) with Thiocyanate Ion in Aqueous PEG Solutions

The spectrophotometric study on the formation reactions of Co(II) thiocyanate complexes was carried out using the solutions containing Co(II) at 2.0×10^{-4} mol L^{-1}, PEG4000 (Mw (molecular weight) = 2700–3400) at 50%(w/v), and thiocyanate ion at various concentrations. The ionic strength and pH of the solutions were adjusted to 0.4 mol L^{-1} and 3 by the addition of sodium perchlorate and nitric acid, respectively, as it has been revealed that any complex is not formed between Co(II) and either perchlorate or nitrate ion [30].

Figure 2.2 shows the variation of the absorption spectra of the PEG4000 solution containing Co(II) with the addition of thiocyanate ion. The absorbance at 624 nm increases appreciably with increasing the concentration of thiocyanate ion, C_{NCS}. The effect of the concentration of PEG4000 on the absorbance at 624 nm is shown in Fig. 2.3, which indicates that the absorbance increases with an increase in the PEG concentration in the solution. These results demonstrate that the formation reaction of tetrahedral $Co(NCS)_4{}^{2-}$ complex is enhanced by the presence of PEG molecules in aqueous solutions [25–28].

It is possible to determine the molar absorption coefficients and formation constants for the Co(II) thiocyanate complexes from the spectrophotometric data obtained. The overall formation constant for the n-coordinated thiocyanate complex, β_n, is defined by Eq. (2.4):

$$Co^{2+} + nNCS^- \rightleftarrows Co(NCS)_n^{(2-n)+} \quad (n = 1 - 4)$$

$$\beta_n = \frac{[Co(NCS)_n^{(2-n)+}]}{[Co^{2+}][NCS^-]^n} \qquad (n = 1 - 4) \tag{2.4}$$

The absorbance, A, of a given system containing free Co(II) ion ($Co(H_2O)_6{}^{2+}$) and n complex species is given by

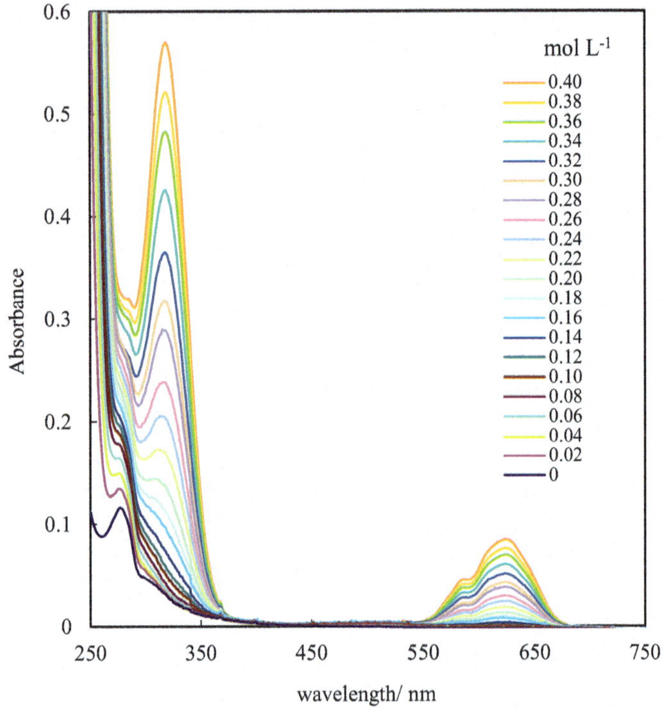

Fig. 2.2 Absorption spectra of Co(II) in 50%(w/v) PEG4000 solution containing NaNCS. Temperature: 25 °C; ionic strength: 0.4 mol L^{-1}; total concentration of Co(II): 2.0×10^{-4} mol L^{-1}; total concentration of NCS^-: 0–0.40 mol L^{-1} (reproduced from Ref. [29] by permission of Elsevier)

$$A = \varepsilon_0[Co^{2+}] + \sum \varepsilon_n \beta_n [Co^{2+}][NCS^-]^n \quad (2.5)$$

where ε_0 and ε_n are the molar absorption coefficients of $Co(H_2O)_6{}^{2+}$ and $Co(NCS)_n{}^{(2-n)+}$ complexes in the solution. The absorbance of thiocyanate ion can be neglected since it does not show absorption in the visible region.

The material balance equations for Co(II) and thiocyanate ion are given in Eqs. (2.6) and (2.7), respectively:

$$C_{Co} = [Co^{2+}] + \sum \beta_n [Co^{2+}][NCS^-]^n \quad (2.6)$$

$$C_{NCS} = [NCS^-] + \sum n\beta_n [Co^{2+}][NCS^-]^n \quad (2.7)$$

where C_{Co} represents the total concentration of Co(II). It can be assumed that the coordination of hydroxide ion to Co(II) is negligible since all the experiments were carried out at pH 3.

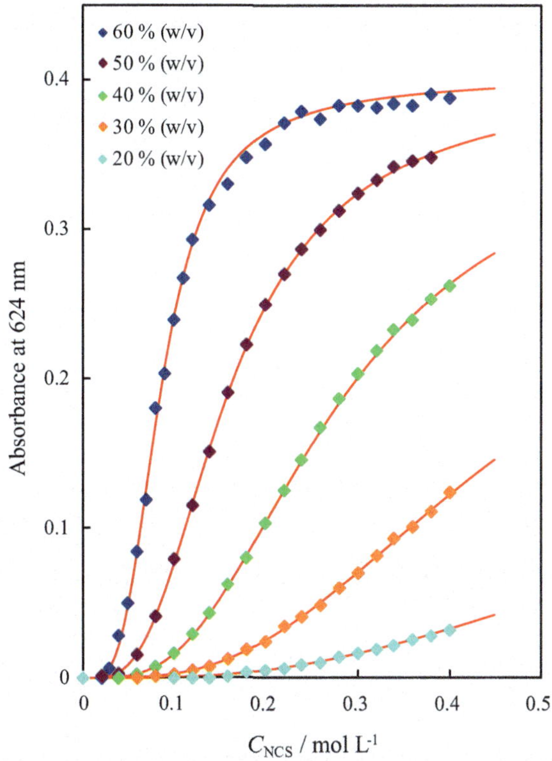

Fig. 2.3 Dependence of the absorbance at 624 nm of 20–60%(w/v) PEG4000 solutions on C_{NCS}. Temperature: 25 °C; ionic strength: 0.4 mol L^{-1}; total concentration of Co(II): 2.0 × 10^{-4} mol L^{-1}. Solid lines represent the best curve fits to the experimental data (reproduced from Ref. [29] by permission of Elsevier)

The solid lines shown in Fig. 2.3 represent the best curve fits to the experimental data obtained using Eqs. (2.4)–(2.7). The apparent overall formation constants, log β_n, obtained for the solutions containing 0–60%(w/v) PEG4000 are listed in Table 2.1. In Fig. 2.4, the molar absorption coefficient spectra extracted for $Co(H_2O)_6{}^{2+}$ and $Co(NCS)_n{}^{(2-n)+}$ ($n = 1$–4) in the range of 400–700 nm are shown, which indicates that the absorption in the visible region is attributed to the formation of $Co(NCS)_4{}^{2-}$. The ε_4 value at 624 nm (2040 L mol^{-1} cm^{-1}) is in good agreement with those of the complex in DMF [18] and micellar solutions of Triton X [25–28], suggesting that an aqueous PEG solution gives a solvent environment similar to those of aprotic solvents and micelles of nonionic surfactants.

Figure 2.5 shows the distribution of the Co(II) thiocyanato complexes in PEG4000 solutions. As can be seen in Fig. 2.5, the increase in the concentration of PEG4000 promotes the formation of $Co(NCS)_4{}^{2-}$. Figure 2.6 represents the dependences of log β_n values on the concentration of PEG4000 in solution. It can be seen that log β_4 increases with an increase in the concentration of PEG, while the log β values for the other thiocyanato complexes are nearly constant independent of the PEG concentration. These results imply that an aqueous PEG solution has a heterogeneous structure and the formation of tetrahedral $Co(NCS)_4{}^{2-}$ complex proceeds in the vicinity of PEG molecules.

Table 2.1 Calculated overall formation constants of $Co(NCS)_n^{(2-n)+}$ complexes (n = 1–4) in aqueous solutions containing 0–60%(w/v) PEG4000. Temperature: 25 °C; ionic strength: 0.4 mol L^{-1} (reproduced from Ref. [29] by permission of Elsevier)

PEG % (w/v)	log β_1	log β_2	log β_3	log β_4
0[a]	0.95	1.60	1.78	−0.30
20	0.83	1.08	1.42	1.41
30	0.84	0.85	1.55	2.09
40	1.15	1.39	1.92	3.08
50	0.98	1.78	2.15	3.78
60	0.86	1.43	2.53	4.54

[a]Literature data [31]

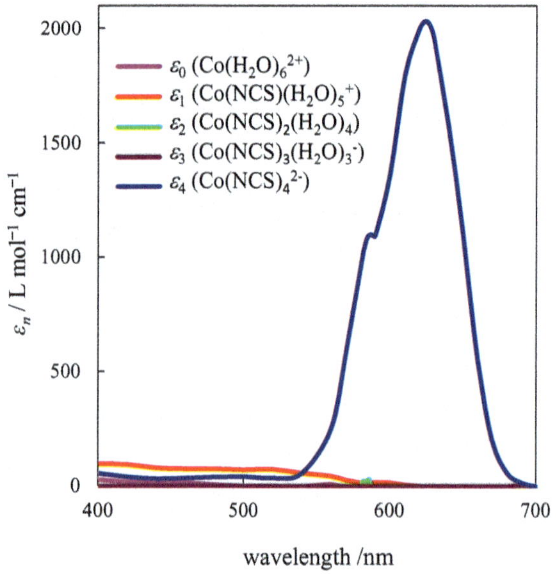

Fig. 2.4 Molar absorption coefficient spectra of $Co(NCS)_n^{(2-n)+}$ complexes extracted by the curve fitting. Temperature: 25 °C; ionic strength: 0.4 mol L^{-1} (reproduced from Ref. [29] by permission of Elsevier)

2.3 Effect of Terminal Functional Groups of Ethylene Oxide Chains on the Formation of Co(II) Thiocyanate Complexes

The electronic spectra for Co(II) thiocyanate complexes in the PEG solutions of varying ethylene oxide chain lengths show similar profiles. However, the absorbance at 624 nm increases with an increasing number of ethylene oxide groups as shown in Fig. 2.7; ethylene glycol (EG) < diethylene glycol (DEG) < PEG400 (Mw = 380–420) < PEG1000 (Mw = 950–1050) < PEG4000 < PEG6000 (Mw = 7000–10,200). This suggests that the chain length of ethylene oxide groups has a decisive role on the formation of the tetrahedral $Co(NCS)_4^{2-}$ complex. Umebayashi et al. also reported

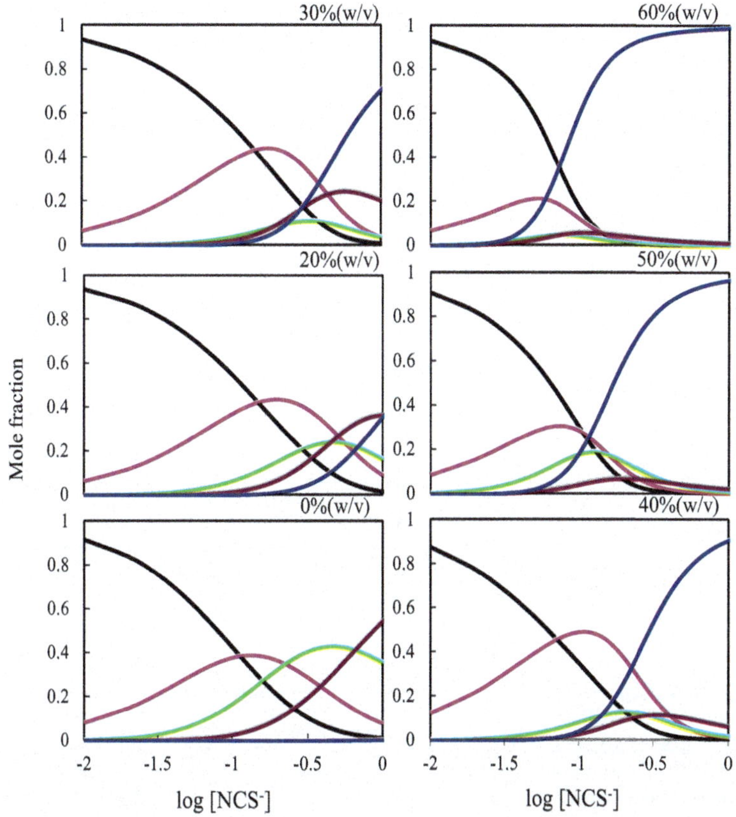

Fig. 2.5 Distribution of Co(II) thiocyanate complexes in 0–60%(w/v) aqueous PEG4000 solutions. Temperature: 25 °C; ionic strength: 0.4 mol L^{-1}; total concentration of Co(II): 2.0×10^{-4} mol L^{-1}

that the formation of $Co(NCS)_4{}^{2-}$ in the micellar solutions of Triton X is enhanced with increasing number of ethylene oxide groups of the nonionic surfactants [28].

The apparent overall formation constants of $Co(NCS)(H_2O)_5{}^+$, $Co(NCS)_2(H_2O)_4$, $Co(NCS)_3(H_2O)_3{}^-$, and $Co(NCS)_4{}^{2-}$ calculated from the spectrometric data with Eqs. (2.4)–(2.7) are tabulated in Table 2.2. As seen in Table 2.2, the formation constant of the tetrahedral $Co(NCS)_4{}^{2-}$ increases with an increase in the ethylene oxide chain length of a PEG molecule, while those of the octahedral complexes do not show any dependence on the molecular weight of PEG.

EG, DEG, and PEGs used have hydroxyl groups at both ends of the ethylene oxide chain, which means that an increase in the number of ethylene oxide groups in the molecule corresponds to a decrease in the concentration of terminal hydroxyl groups in solution when the weight percent of the compound is constant. As it has been reported that the aggregation structure strongly depends on the terminal groups of PEG [32], the polarity of the terminal group may have a significant effect on

Fig. 2.6 Apparent overall formation constants of Co(II) thiocyanate complexes in PEG4000 solutions of varying concentration (0–60%(w/v)). Temperature: 25 °C; ionic strength: 0.4 mol L^{-1} (reproduced from Ref. [29] by permission of Elsevier)

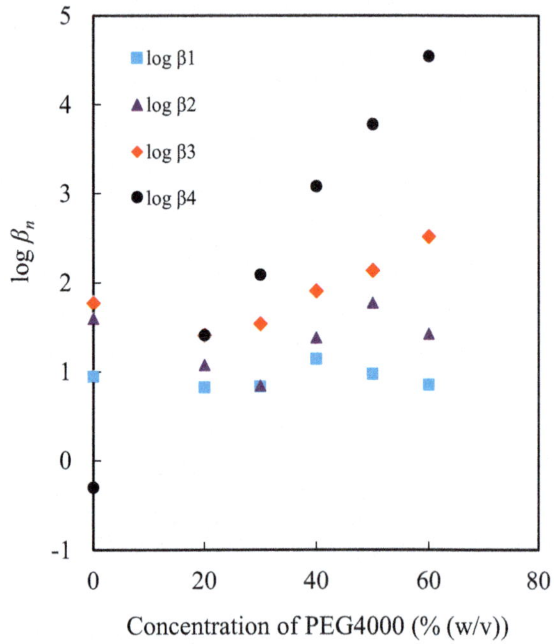

Fig. 2.7 Dependence of the absorbance for $Co(NCS)_4^{2-}$ at 624 nm in 50%(w/v) EG, DEG, and PEG solutions on thiocyanate concentration. Temperature: 25 °C; ionic strength: 0.4 mol L^{-1}; total concentration of Co(II): 2.0 × 10^{-4} mol L^{-1}. Solid lines represent the best curve fits to the experimental data (reproduced from Ref. [29] by permission of Elsevier)

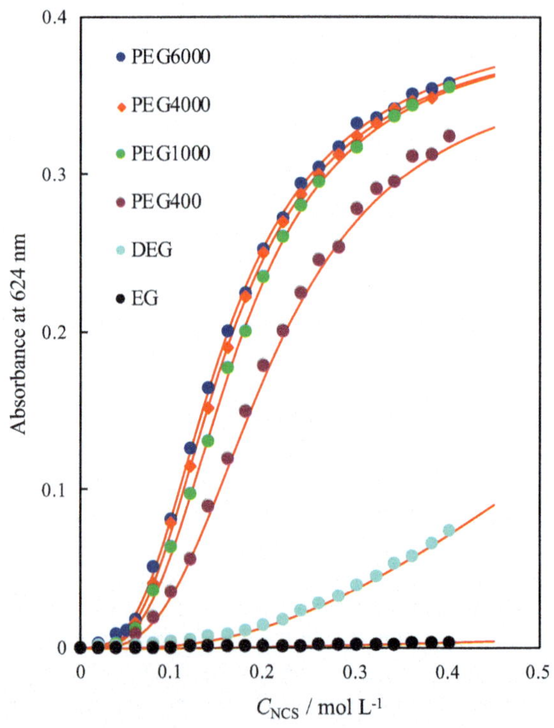

Table 2.2 Apparent overall formation constants of $Co(NCS)_n{}^{(2-n)+}$ complexes ($n = 1$–4) in solutions containing EG, DEG, PEG400, PEG1000, PEG4000, and PEG6000 at 50%(w/v). Temperature: 25 °C; ionic strength: 0.4 mol L^{-1} (reproduced from Ref. [29] by permission of Elsevier)

	EG	DEG	PEG400	PEG1000	PEG4000	PEG6000
$\log \beta_1$	0.64	1.15	1.24	1.31	0.98	1.20
$\log \beta_2$	1.33	1.34	1.26	1.11	1.78	1.68
$\log \beta_3$	2.11	1.36	2.11	2.35	2.15	2.25
$\log \beta_4$	0.65	2.00	3.49	3.77	3.78	3.91

the complexation reaction through the change in the hydration structure of PEG molecules in aqueous solutions.

The dependences of the absorbance of $Co(NCS)_4{}^{2-}$ at 624 nm on thiocyanate concentration for polyethylene glycol monomethyl ether (MPEG350, Mw = 350) and ethylene glycol dimethyl ether (EGDME) solutions are compared with those for PEG400, PEG1000, and PEG4000 solutions in Fig. 2.8. MPEG350 is a PEG in which one of the terminal hydroxyl groups of ethylene oxide chain is methylated, while EGDME has methoxy groups at both ends of ethylene oxide. As shown in Fig. 2.8, the absorbance for the MPEG350 solution is approximately the same as those observed for PEG1000 and PEG4000 solutions despite MPEG350 having shorter ethylene oxide chain than PEG400. On the other hand, the EGDME solution exhibits even stronger absorption, which is greater than that of all the PEG solutions. These results indicate that methylation of terminal hydroxyl groups of a PEG molecule enhances the formation of $Co(NCS)_4{}^{2-}$ in the solutions.

The effect of terminal hydroxyl groups on the formation of $Co(NCS)_n{}^{(2-n)+}$ was evaluated by varying the concentration of terminal hydroxyl groups in solutions, [OH], using EG, DEG, and PEGs of different molecular weights while keeping the concentration of each compound at 50%(w/v) and measuring the apparent overall formation constants of the complexes. The concentration of hydroxyl groups in a PEG solution can be calculated as

$$[\mathrm{OH}] = \frac{2C_{\mathrm{PEG}}}{M_{\mathrm{W,av}}} \tag{2.8}$$

where C_{PEG} and $M_{\mathrm{w,av}}$ denote the concentration of PEG in units of g L^{-1} and the average molecular weight calculated as a median of the molecular weight range of the PEG given by the manufacturer. Figure 2.9 shows the plots of $\log \beta_n$ versus [OH]. As seen in Fig. 2.9, the $\log \beta_4$ value decreases with an increase in [OH], while the other formation constant values for the octahedral complexes do not show clear dependence on [OH], indicating that the formation of tetrahedral $Co(NCS)_4{}^{2-}$ complex is inhibited by terminal hydroxyl groups of PEG molecules.

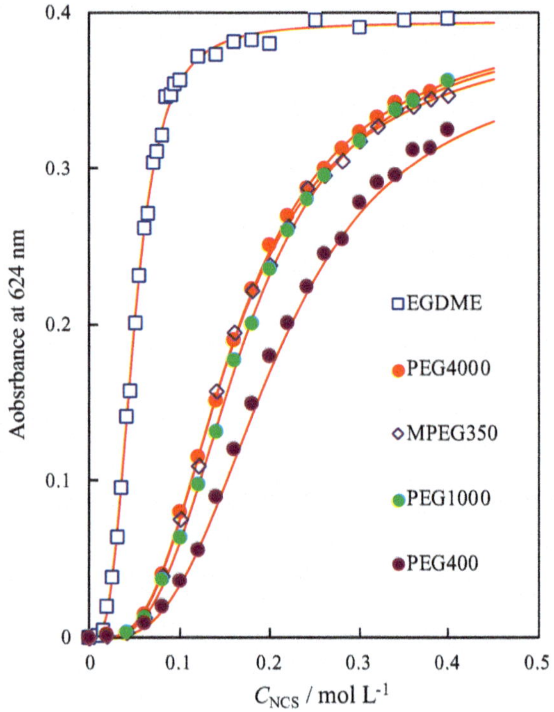

Fig. 2.8 Dependence of the absorbance for $Co(NCS)_4^{2-}$ at 624 nm in EGDME, MPEG350, PEG400, PEG1000, and PEG4000 solutions on thiocyanate concentration. Temperature: 25 °C; ionic strength: 0.4 mol L^{-1}; total concentration of Co(II): 2.0×10^{-4} mol L^{-1}; concentration of EGDME, MPEG350, and PEGs: 50%(w/v). Solid lines represent the best curve fits to the experimental data (reproduced from Ref. [29] by permission of Elsevier)

2.4 A Micro-Heterogeneous Structure Model of PEG–Water Mixtures

From the experimental results described above, it can be concluded that specific solute partitioning behavior in PEG-based ATPS is the result of the compound effect of hydration structures of ethylene oxide (EO) units and terminal hydroxyl groups (OH) of PEG in addition to the bulk water phase (bulk) in the upper and lower phases. The clustering or aggregation phenomenon has been observed for PEG molecules in water [15–17, 32–37]. Zaslavsky et al. predicted that phase separation in mixtures of the aqueous solutions of two different polymers or a polymer and an inorganic salt occurs due to the formation of dissimilar micro-domains in the mixture, and showed using DLS that polymer aggregation occurs at concentrations significantly lower than that corresponding to visual macroscopic phase separation [16]. Figure 2.10 shows size distribution of aggregates of PEG8000 for diluted solutions of the ATPS consisting of 11.1 wt% PEG8000 and 6.3 wt% Na_2SO_4 compared with the solution of PEG8000 alone. They describe that in their DLS experiments, the dispersant is a solution of PEG and salt, and the "particles" observed are aggregates of PEG present in the dispersant. Phase separation occurs in the system with 90% ATPS, while dilution systems at 80% or below are visually homogeneous (the phase diagram of PEG8000/Na_2SO_4 system is shown in Fig. 2.11). Therefore, these DLS results

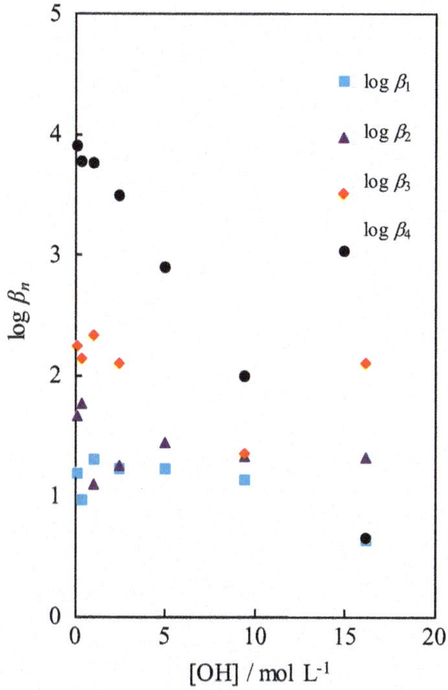

Fig. 2.9 Dependence of the apparent overall formation constants of Co(II) thiocyanate complexes on the concentration of terminal hydroxyl groups, [OH], in aqueous solutions of EG, DEG, PEG400, PEG1000, PEG4000, and PEG6000. Temperature: 25 °C; ionic strength: 0.4 mol L^{-1}; concentration of EG, DEG, and PEGs: 50%(w/v) (reproduced from Ref. [29] by permission of Elsevier)

indicate that PEG8000 forms aggregates in water and the size of the aggregates increases with an increase in concentrations of PEG and Na_2SO_4 reaching over 100 nm. It is thus reasonable to assume that the local hydration structures of EO and OH groups as well as the bulk water phase in the PEG-rich phase have sizes which can be regarded as individual independent phases and the complexation reactions of Co(II) with thiocyanate ion proceed in these phases independently of one another [38, 39].

Based on the assumption described above, the intrinsic overall formation constants, $\beta_{n,i}$, of the complexes $Co(NCS)_n^{(2-n)+}$ in the three phases (EO, OH, and bulk phases) and the volumes of the phases were estimated by analyzing the spectrophotometric data. $\beta_{n,i}$ of the complex $Co(NCS)_n^{(2-n)+}$ in the phase i (i = EO, OH, or bulk) is defined by

$$\beta_{n,i} = \frac{\left[Co(NCS)_n^{(2-n)+}\right]_i}{\left[Co^{2+}\right]_i [NCS^-]_i^n} \tag{2.9}$$

The partition coefficients of $Co(H_2O)_6^{2+}$ and thiocyanate ion, $K_{Co,\,i}$ and $K_{NCS,\,i}$, from the bulk water phase to the EO and OH phases are represented by Eqs. (2.10) and (2.11), respectively.

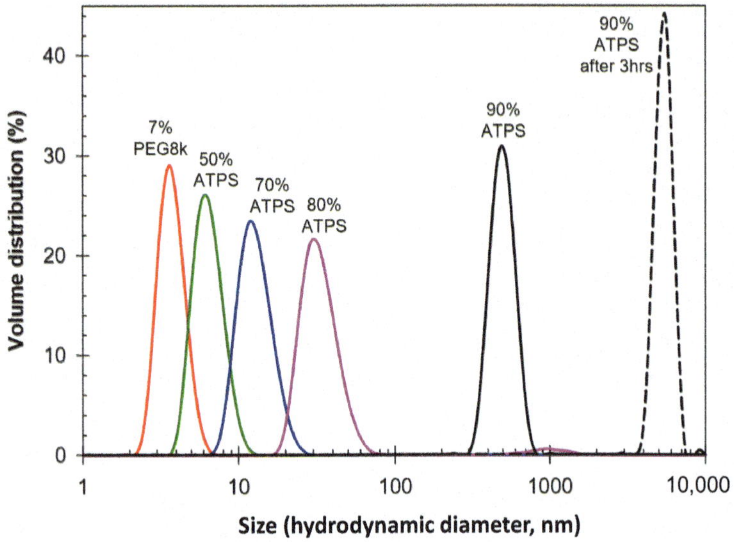

Fig. 2.10 Size distribution for 7% PEG8000 (red) compared to various dilutions of an ATPS consisting of 11.1% PEG8000 and 6.3% Na_2SO_4. Mixed dilutions of the ATPS correspond to 5.55% PEG8000 and 3.15% Na_2SO_4 (green), 7.77% PEG8000 and 4.41% Na_2SO_4 (blue), 8.88% PEG8000 and 5.04% Na_2SO_4 (pink), 10% PEG8000 and 5.7% Na_2SO_4 freshly mixed (black), and after standing for 3 h (black dashed). All solutions prepared in phosphate buffer (pH 7.4); concentrations in wt% (reproduced from Ref. [16])

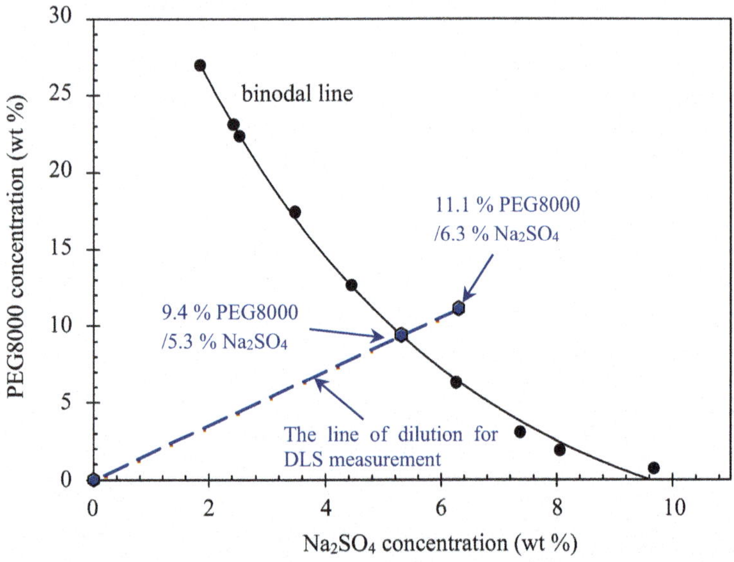

Fig. 2.11 Phase diagram of PEG8000/Na_2SO_4 ATPS (reproduced from Ref. [16])

$$K_{\mathrm{Co},\,i} = \frac{[\mathrm{Co}^{2+}]_i}{[\mathrm{Co}^{2+}]_{\mathrm{bulk}}} \tag{2.10}$$

$$K_{\mathrm{NCS,i}} = \frac{[\mathrm{NCS}^-]_i}{[\mathrm{NCS}^-]_{\mathrm{bulk}}} \tag{2.11}$$

The concentrations of $Co(H_2O)_6{}^{2+}$ and thiocyanate ion in the EO, OH, and bulk water phases are given by the following equations:

$$[\mathrm{Co}^{2+}]_{\mathrm{EO}} = \frac{VK_{\mathrm{Co,\,EO}}[\mathrm{Co}^{2+}]_{\mathrm{T}}}{V + V_{\mathrm{EO}}(K_{\mathrm{Co,\,EO}} - 1) + V_{\mathrm{OH}}(K_{\mathrm{Co,\,OH}} - 1)} \tag{2.12}$$

$$[\mathrm{Co}^{2+}]_{\mathrm{OH}} = \frac{VK_{\mathrm{Co,\,OH}}[\mathrm{Co}^{2+}]_{\mathrm{T}}}{V + V_{\mathrm{EO}}(K_{\mathrm{Co,\,EO}} - 1) + V_{\mathrm{OH}}(K_{\mathrm{Co,\,OH}} - 1)} \tag{2.13}$$

$$[\mathrm{Co}^{2+}]_{\mathrm{bulk}} = \frac{V[\mathrm{Co}^{2+}]_{\mathrm{T}}}{V + V_{\mathrm{EO}}(K_{\mathrm{Co,\,EO}} - 1) + V_{\mathrm{OH}}(K_{\mathrm{Co,\,OH}} - 1)} \tag{2.14}$$

$$[\mathrm{NCS}^-]_{\mathrm{EO}} = \frac{VK_{\mathrm{NCS,\,EO}}[\mathrm{NCS}^-]_{\mathrm{T}}}{V + V_{\mathrm{EO}}(K_{\mathrm{NCS,\,EO}} - 1) + V_{\mathrm{OH}}(K_{\mathrm{NCS\,OH}} - 1)} \tag{2.15}$$

$$[\mathrm{NCS}^-]_{\mathrm{OH}} = \frac{VK_{\mathrm{NCS,\,OH}}[\mathrm{NCS}^-]_{\mathrm{T}}}{V + V_{\mathrm{EO}}(K_{\mathrm{NCS,\,EO}} - 1) + V_{\mathrm{OH}}(K_{\mathrm{NCS\,OH}} - 1)} \tag{2.16}$$

$$[\mathrm{NCS}^-]_{\mathrm{bulk}} = \frac{V[\mathrm{NCS}^-]_{\mathrm{T}}}{V + V_{\mathrm{EO}}(K_{\mathrm{NCS,\,EO}} - 1) + V_{\mathrm{OH}}(K_{\mathrm{NCS\,OH}} - 1)} \tag{2.17}$$

where $[Co^{2+}]_T$ and $[NCS^-]_T$ are the total concentrations of $Co(H_2O)_6{}^{2+}$ and NCS^- ions in the PEG solution, respectively. V is the volume of the solution and V_{EO}, V_{OH}, and V_{bulk} are the volumes of the phases of EO, OH, and the bulk water, respectively. V is given by the sum of V_{EO}, V_{OH}, and V_{bulk} as

$$V = V_{\mathrm{EO}} + V_{\mathrm{OH}} + V_{\mathrm{bulk}} \tag{2.18}$$

$[Co^{2+}]_T$ and $[NCS^-]_T$ are related to C_{Co} and C_{NCS} by the following mass balance equations through the apparent overall formation constants as follows:

$$C_{\mathrm{Co}} = [\mathrm{Co}^{2+}]_{\mathrm{T}} + \sum \beta_n[\mathrm{Co}^{2+}]_{\mathrm{T}}[\mathrm{NCS}^-]_{\mathrm{T}}^n \tag{2.19}$$

$$C_{\mathrm{NCS}} = [\mathrm{NCS}^-]_{\mathrm{T}} + \sum n\beta_n[\mathrm{Co}^{2+}]_{\mathrm{T}}[\mathrm{NCS}^-]_{\mathrm{T}}^n \tag{2.20}$$

Since only tetrahedral thiocyanate complex shows absorption at 624 nm, the absorbance of the solution at this wavelength is given by

$$A = \sum \left(\frac{V_i}{V}\right) \varepsilon_4 [\mathrm{Co(NCS)_4^{2-}}]_i = \sum \left(\frac{V_i}{V}\right) \varepsilon_4 \beta_{4,i} [\mathrm{Co^{2+}}]_i [\mathrm{NCS^-}]_i^4$$
$$(i = \mathrm{EO,\ OH,\ or\ bulk}) \quad (2.21)$$

The values of $K_{\mathrm{Co,\,EO}}$, $K_{\mathrm{Co,\,OH}}$, $K_{\mathrm{NCS,\,EO}}$, $K_{\mathrm{NCS,\,OH}}$, $\beta_{4,\,\mathrm{EO}}$, $\beta_{4,\,\mathrm{OH}}$, V_{EO}/V, and V_{OH}/V calculated by the curve fitting of the absorption spectra for the solutions of EG, DEG, PEGs, and MPEG350 with ethylene oxide concentrations of 10 mol L^{-1} and 12 mol L^{-1} are tabulated in Table 2.3. The concentration of ethylene oxide units in a PEG solution, [EO], was calculated as

$$[\mathrm{EO}] = \frac{(M_{\mathrm{W,av}} - 18)C_{\mathrm{PEG}}}{44M_{\mathrm{W,av}}} \quad (2.22)$$

The results shown in Table 2.3 indicate that the spectrophotometric data obtained for the PEG solutions can be satisfactorily explained by the three-phase (EO, OH, and bulk water) model. The log $\beta_{4,\mathrm{EO}}$ values for all the systems studied are in good agreement with one another, whereas the log $\beta_{4,\mathrm{OH}}$ value is very small in every system, indicating that the formation of $\mathrm{Co(NCS)_4}^{2-}$ is negligible in the OH phase.

The $K_{\mathrm{Co,\,EO}}$ and $K_{\mathrm{NCS,\,EO}}$ values are almost equal to the partition coefficients of free Co^{2+} and NCS^- ions in PEG/Na_2SO_4 ATPS in which the $\Delta\mathrm{EO}_{,\mathrm{m}}$ are 10 mol L^{-1}, 0.03 and 12, respectively [20]. It is interesting to note that the value of $K_{\mathrm{Co,\,OH}}$ is larger and closer to unity than the $K_{\mathrm{Co,\,EO}}$ value, while $K_{\mathrm{NCS,\,OH}}$ is smaller and also

Table 2.3 Partition coefficients of free Co^{2+} and thiocyanate ions ($K_{\mathrm{Co,\,EO}}$, $K_{\mathrm{Co,\,OH}}$, $K_{\mathrm{NCS,\,EO}}$, and $K_{\mathrm{NCS,\,OH}}$); the intrinsic overall formation constants of $\mathrm{Co(NCS)_4}^{2-}$ in the EO and OH phases (log $\beta_{4,\,\mathrm{EO}}$ and log $\beta_{4,\,\mathrm{OH}}$); and the values of V_{EO}/V and V_{OH}/V for PEGs, EG, DEG, and MPEG350 solutions containing 10 mol L^{-1} and 12 mol L^{-1} EO units. Temperature: 25 °C; ionic strength: 0.4 mol L^{-1} (reproduced from Ref. [29] by permission of Elsevier)

	[EO]/ mol L^{-1}	[OH]/ mol L^{-1}	$K_{\mathrm{Co,EO}}$	$K_{\mathrm{Co,OH}}$	$K_{\mathrm{NCS,EO}}$	$K_{\mathrm{NCS,OH}}$	log $\beta_{4,\mathrm{EO}}$	log $\beta_{4,\mathrm{OH}}$	V_{EO}/V	V_{OH}/V
PEG6000	10	0.1	0.01	0.04	9.6	2.7	4.00	<−2	0.95	<0.01
PEG4000	10	0.3	0.01	0.01	16	3.6	4.00	<−2	0.94	0.01
PEG1000	10	0.9	0.01	0.04	9.3	3.4	4.00	<−2	0.89	0.03
PEG400	10	2.3	0.01	0.02	9.7	3.1	3.99	<−2	0.80	0.07
DEG	10	10.0	0.01	0.08	8.0	4.0	3.99	−0.3	<0.01	0.30
EG	10	20.0	0.01	0.04	5.0	3.2	3.99	−0.5	<0.01	0.55
MPEG350	10	1.4	0.01	0.02	13	3.6	4.03	<−2	0.95	0.03
PEG4000	12	0.3	0.01	0.02	9.9	3.6	4.00	<−2	0.99	0.02
PEG1000	12	1.1	0.02	0.01	25	1.2	4.03	<− 2	0.95	0.05
PEG400	12	2.8	0.01	<0.01	12	0.5	4.08	<−2	0.85	0.12
DEG	12	12.0	0.02	0.04	12	4.3	4.00	<− 2	<0.01	0.31
EG	12	24.0	0.01	0.05	8.4	4.3	3.99	−0.9	<0.01	0.72

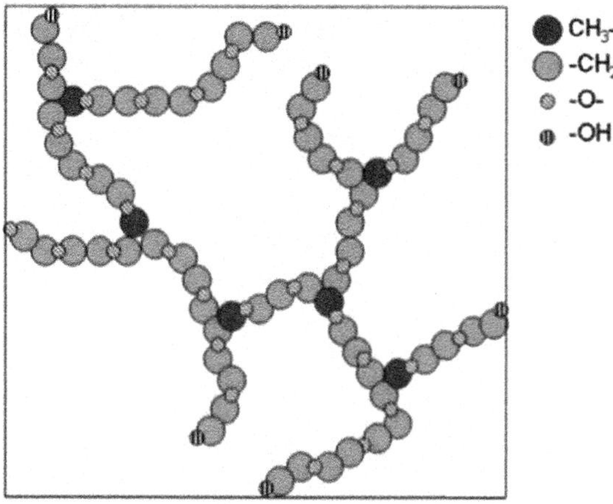

Fig. 2.12 Schematic rendering (not to scale and not realistic) of the PEO/d-water clusters in the case with different chain ends (–OCH3 and –OH) (reproduced from Ref. [32] by permission of The American Chemical Society)

closer to unity than $K_{NCS, EO}$. This suggests that the OH pseudo-phase has similar solute dissolution properties to those of the bulk water.

The V_{EO} value increases for higher molecular weight PEGs, while the V_{OH} value increases for lower molecular weight PEGs at identical [EO]. This corresponds to the change in the relative concentrations of EO and OH groups in the PEG solution. In EG and DEG solutions, the V_{EO} values detected are extremely small, suggesting that strong hydration of terminal hydroxyl groups interferes with the formation of hydrophobic hydration structure around ethylene oxide groups of these small molecules. On the other hand, the V_{EO} value for the MPEG350 solution is comparable to that for the solution of PEG6000 even though the V_{OH} value for the MPEG350 solution is much larger than the value for PEG6000 solution. These results may be explained by the strong hydrophobic hydration of the terminal methyl groups or specific aggregation structures formed by the hydrophobic interaction between ethylene oxide chains and methyl groups as suggested by Hammouda et al. (Fig. 2.12) [32].

The volume fraction of ethylene oxide groups in the solution containing 12 mol L^{-1} EO is calculated to be 0.44 from the molar volume of ethylene oxide group (37 mL mol^{-1}) [40]. However, the V_{EO}/V values obtained for the solutions of PEG400, PEG1000, and PEG4000 are much larger than 0.44 as shown in Table 2.3. This means that most of the water is hydrated with ethylene oxide groups of PEG molecules and forms the EO pseudo-phase, resulting in a decrease in the volume fraction of the bulk water in solution.

It can be concluded from the results presented above that the EO pseudo-phase is formed by the involvement of hydrophobic ethylene groups and long ethylene oxide chains of PEG molecules are essential for the formation of the EO phase. Privat et al. investigated the mechanism of aggregation of PEG molecules by fluorescence spectroscopy and surface tension measurements and showed that a PEG molecule forms a helical structure with inner water molecules hydrated to ether oxygen and ethylene groups set outside the helix [15, 41]. Sun et al. studied the distribution of $HCrO_4^-$ and CrO_4^{2-} ions around the PEG aggregates by molecular dynamics simulation [42]. Figure 2.13 shows the radial density profiles of $HCrO_4^-$ and CrO_4^{2-} ions as well as PEG, H_2O, Na^+, and SO_4^{2-} around the center of PEG aggregates in aqueous solution of Na_2SO_4. As can be seen in Fig. 2.13, poorly hydrated $HCrO_4^-$ ions distribute in the PEG aggregates, while strongly hydrated CrO_4^{2-} and SO_4^{2-} are excluded from the PEG aggregates and distribute in the bulk aqueous solution. A snapshot in the inner of PEG aggregates is presented in Fig. 2.14. As reported in many studies, sodium ions are surrounded by ether oxygen atoms of PEG molecules to form the complex in a similar manner to the cyclic one (crown ether) [26, 33, 43–45]. On the other hand, $HCrO_4^-$ is located outside ethylene groups. $Co(NCS)_4^{2-}$ has a weak hydrogen bonding ability and the hydrogen-bonded network structure of water around the complex is reported to be broken [28]. From these experimental and MD simulation results, it is considered that the formation of $Co(NCS)_4^{2-}$ takes place at a hydrophobic area in the EO phase out of the helix in PEG aggregates. NCS^- ion is also known to be one of the most chaotropic ions [46].Therefore, the large $K_{NCS,EO}$ values can also be attributed to its weak hydration, resulting in the distribution of the anion into the EO phase.

Figure 2.15 shows a schematic picture of the distribution of $Co(H_2O)_6^{2+}$, $Co(NCS)(H_2O)_5^+$, and $Co(NCS)_4^{2-}$ in the estimated structure of aqueous PEG solutions [29]. The EO phase formed by ethylene oxide groups of PEG plays a decisive role in the partitioning of solute molecules having hydrophobic groups and chaotropic anions into the PEG-rich phase in ATPS.

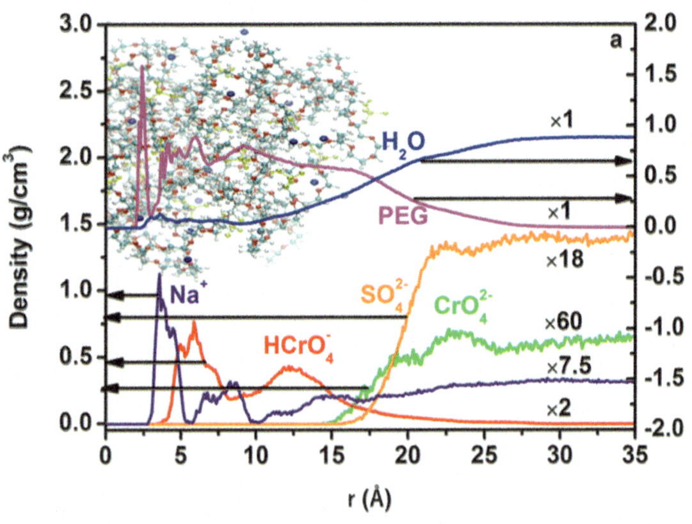

Fig. 2.13 Radial density profiles of $HCrO_4^-$, CrO_4^{2-}, PEG, H_2O, Na^+, and SO_4^{2-} around the center of PEG aggregates (reproduced from Ref. [42] by permission of The American Chemical Society)

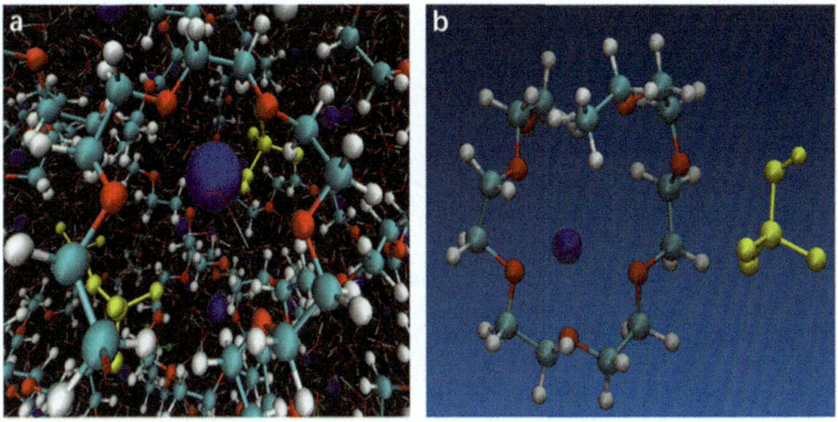

Fig. 2.14 Snapshots of the inner of PEG aggregates (**a**) and PEG molecules and $HCrO_4^-$ ion (**b**). Carbon, hydrogen, and oxygen atoms in PEG molecules are represented by cyan, white, and red spheres, respectively. $HCrO_4^-$ is represented by yellow spheres. Na^+ is represented by blue spheres (reproduced from Ref. [42] by permission of The American Chemical Society)

2.5 Thermal Phase Transition Behavior of Water in Aqueous PEG Solutions

In aqueous solutions containing solute compounds, water exhibits thermodynamic properties different from those in the pure bulk liquid phase due to the solute–water interactions. It has been known that DSC can give quantitative information about water in a different state than the bulk water in the solutions or at the surfaces of solid materials as freezable-bound water and nonfreezing water; the former is water which exhibits melting and crystallization temperatures lower than that of the bulk water and the latter is water that does not show a phase transition in the normal temperature range.

Water in PEG–water mixtures also shows specific phase transition behavior in DSC measurements. Figure 2.16 shows the DSC heating curves obtained for the aqueous solutions of PEG4000 with various total water content, w_t. Only one endothermic peak corresponding to the melting of water was observed when $w_t < 0.5$, while two peaks were observed when $w_t > 0.5$. The amounts of the bulk water, of which the melting temperature is 273 K, w_{bulk}, and of freezable-bound water, w_{fb}, expressed in g g^{-1} dry polymer, can be estimated by the following equations, respectively.

$$w_{bulk} = Q_H/\Delta H W_g \tag{2.23}$$

$$w_{fb} = Q_L/\Delta H W_g \tag{2.24}$$

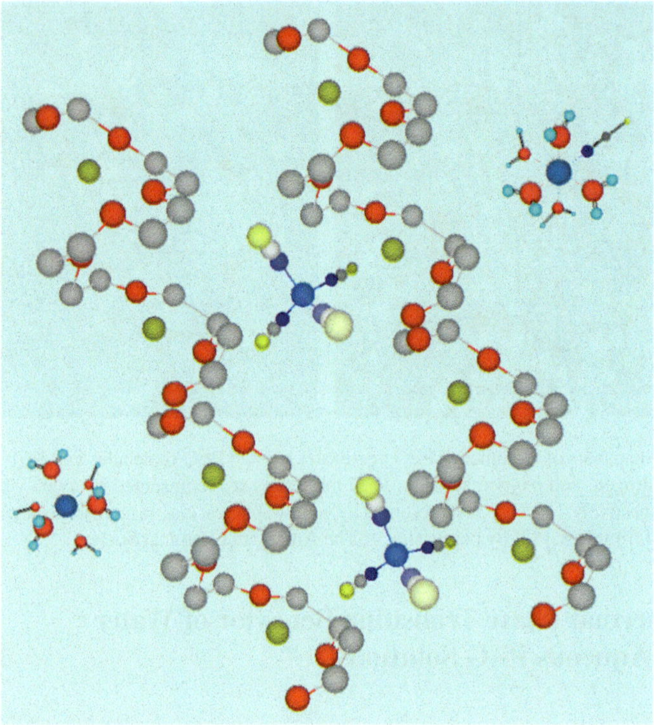

Fig. 2.15 Schematic representation of the distribution of $Co(H_2O)_6{}^{2+}$, $Co(NCS)(H_2O)_5{}^{+}$, and $Co(NCS)_4{}^{2-}$ in the estimated structure of an aqueous PEG solution. Atoms are colored as follows: C, gray; O, red; Na, green; N, blue; S, yellow; Co, light blue (reproduced from Ref. [29] by permission of Elsevier)

where Q_H and Q_L are the heats absorbed in the melting process, which are calculated from areas of the peaks at higher and lower temperatures than 273 K in the DSC heating curve, respectively, ΔH is enthalpy for the melting of water at the corresponding temperatures, and W_g is the weight of the dry PEG4000 [7].

The contents of the bulk water and the freezable-bound water in the PEG–water mixtures depend on the total content of water as shown in Fig. 2.17. The w_{bulk} value decreases with a decrease in w_t, while the w_{fb} value is constant until the bulk water vanishes at $w_t = 1.0$. For the PEG4000 solution, this w_t value corresponds to 12 mol L^{-1} EO. As w_t decreases below 1.0, the w_{fb} value decreases linearly with w_t and disappears at $w_t = 0.1$. These DSC results suggest that there is no bulk water in the aqueous PEG solution of 12 mol L^{-1} EO and the solution consists only of two types of hydration structures detected as freezable-bound water and nonfreezing water and the former occupies 90% of water in the PEG solution, while the latter does 10%. The volume fractions of water to total water for the EO and the OH phases are calculated to be 0.97 and 0.03 from the V_{EO}/V and V_{OH}/V values shown in Table 2.3, respectively. In this calculation, the molar volume of hydroxyl group was assumed

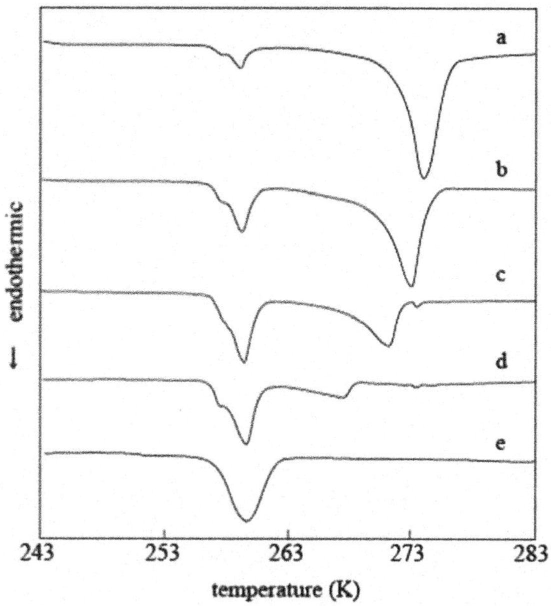

Fig. 2.16 DSC heating curves of aqueous PEG solutions. Total water content, w_t (g/g dry polymer) is a, 8.94; b, 3.93; c, 2.32; d, 1.50; e, 0.44 (reproduced from Ref. [7] by permission of Elsevier)

to be negligibly small because the strong hydrogen bonding with water makes it very small [47].

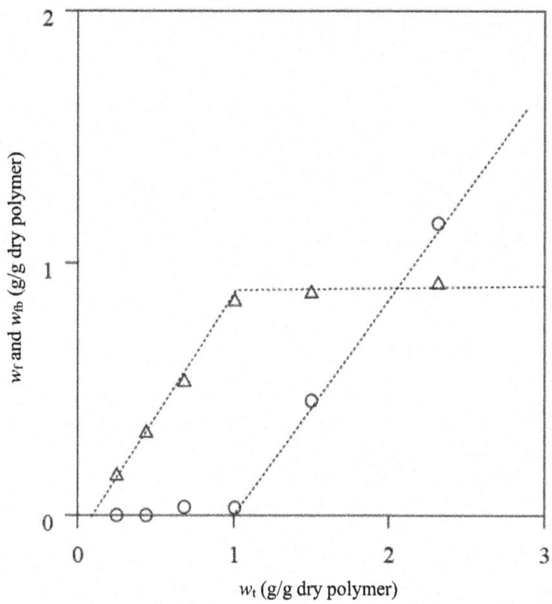

Fig. 2.17 Dependence of w_f and w_{fb} on w_t for aqueous PEG solutions. Symbols: ○ = w_{bulk}; △ = w_{fb} (reproduced from Ref. [7] by permission of Elsevier)

The amounts of freezable-bound water and nonfreezing water obtained by DSC are consistent with those of the EO and OH phase water obtained by spectrophotometry. It is thus presumed that the nonfreezing water in the PEG solution is water hydrogen bonded to terminal hydroxyl groups, while the freezable-bound water is water hydrated to ethylene oxide groups and the hydration shells formed consist of 2.5 and 2.6 water molecules per EO and OH unit, respectively. As the number of water molecules hydrogen-bonded to a PEG molecule is reported to be around 2 per each oxygen of ethylene oxide unit [48–56], this estimation can be regarded as reasonable.

2.6 Conclusions

PEG molecules form aggregates in aqueous solutions and the solution can be regarded as a mixture consisting of the microphases of hydrated EO chains and terminal OH groups of PEG in addition to the bulk water phase. Studies on the complexation equilibria of tetrahedral $Co(NCS)_4{}^{2-}$ complex by spectrophotometry and the thermal phase transition behavior by DSC revealed that most of the water is hydrated to the EO groups of PEG and the bulk water phase virtually disappears in solutions containing PEG at 12 mol L^{-1} EO units or more.

The EO phase plays a predominant role in the distribution of solute molecules and ions into the PEG-rich phase in ATPS. As the tetrahedral $Co(NCS)_4{}^{2-}$ complex has a weak hydrogen bonding ability, the complexation reaction of $Co(NCS)_4{}^{2-}$ proceeds in the EO phase by the involvement of hydrophobic ethylene groups, whereas strongly hydrated structure around terminal OH groups inhibits the formation of the complex. The hydrophobicity, electrostatic property, and hydrogen-bond donor acidity of the PEG-rich phase in PEG/salt ATPS observed by the LSER studies can be attributed to the EO phase, metal ions located on oxygen atoms in the EO units, and the OH phase, respectively. The distribution of solute compounds in PEG-based ATPSs is determined by the difference in the volumes of the EO and OH phases between the upper and lower phases.

References

1. Albertsson PA (1986) Partitioning of cell particles and macromolecules, 3rd edn. Wiley-Interscience, New York
2. Rogers D, Eiteman MA (eds) (1995) Aqueous biphasic separations: biomolecules to metal ions. Plenum Press, New York
3. Saha N, Sarkar B, Sen K (2022) Aqueous biphasic systems: a robust platform for green extraction of biomolecules. J Mol Liq 363:119882
4. Karmakar R, Sen K (2019) Aqueous biphasic extraction of metal ions: an alternative technology for metal regeneration. J Mol Liq 273:231–247

5. Willauer HD, Huddleston JG, Rogers RD (2002) Solute partitioning in aqueous biphasic systems composed of polyethylene glycol and salt: the partitioning of small neutral organic species. Ind Eng Chem Res 41:1892–1904
6. Willauer HD, Huddleston JG, Rogers RD (2002) Solvent properties of aqueous biphasic systems composed of polyethylene glycol and salt characterized by the free energy of transfer of a methylene group between the phases and by a linear solvation energy relationship. Ind Eng Chem Res 41:2591–2601
7. Shibukawa M, Ichikawa R, Baba T, Sakamoto R, Saito S, Oguma K (2008) Separation selectivity of aqueous polyethylene glycol-based separation systems: DSC, LC and aqueous two-phase extraction studies. Polymer 49:4168–4173
8. Huddleston JG, Willauer HD, Rogers RD (2000) Solvatochromic studies in polyethylene glycol–salt aqueous biphasic systems. J Chromatogr B 743:137–149
9. da Silva NR, Ferreira LA, Madeira PP, Teixeira JA, Uversky VN, Zaslavsky BY (2015) Analysis of partitioning of organic compounds and proteins in aqueous polyethylene glycol-sodium sulfate aqueous two-phase systems in terms of solute-solvent interactions. J Chromatogr A 1415:1–10
10. Titus AR, Ferreira LA, Belgovskiy AI, Kooijman EE, Mann EK, Mann JA Jr, Meyer WV, Smart AE, Uversky VN, Zaslavsky BY (2020) Interfacial tension and mechanism of liquid-liquid phase separation in aqueous media. Phys Chem Chem Phys 22:4574–4580
11. da Silva NR, Ferreira LA, Teixeira JA, Uversky VN, Zaslavsky BY (2019) Effects of sodium chloride and sodium perchlorate on properties and partition behavior of solutes in aqueous dextran-polyethylene glycol and polyethylene glycol-sodium sulfate two-phase systems. J Chromatogr A 1583:28–38
12. Taft RW, Abboud JLM, Kamlet MJ, Abraham MH (1985) Linear solvation energy relations. J Sol Chem 14:153–186
13. Abraham MH, Grellier PL (1988) Solvent effects in organic chemistry—recent developments. Can J Chem 66:2673–2686
14. Bagchi S (2015) Solvation in heterogeneous media. In: Birdi KS (ed) Handbook of surface and colloid chemistry, 4th edn. CRC Press, London, pp 461–480
15. Azri A, Giamarchi P, Grohens Y, Olier R, Privat M (2012) Polyethylene glycol aggregates in water formed through hydrophobic helical structures. J Colloid Interf Sci 379:14–19
16. Titus AR, Madeira PP, Ferreira LA, Chernyak VY, Uversky VN, Zaslavsky BY (2022) Mechanism of phase separation in aqueous two-phase systems. Int J Mol Sci 23:14366
17. Paik SP, Ghatak SK, Dey D, Sen K (2012) Poly(ethylene glycol) vesicles: self-assembled site for luminescence generation. Anal Chem 84:7555–7561
18. Ishiguro S, Ozutsumi K (1990) Thermodynamics and structure of isothiocyanato complexes of manganese(II), cobalt(II), and nickel(II) ions in *N,N*-dimethylformamide. Inorg Chem 29:1117–1123
19. Fronæous S, Larsson R (1962) Studies on metal complexes in aqueous solution by infrared spectrophotometry III. Further investigation on thiocyanato complexes. Acta Chem Scand 16:1447–1454
20. Shibukawa M, Nakayama N, Hayashi T, Shibuya D, Endo Y, Kawamura S (2001) Extraction behaviour of metal ions in aqueous polyethylene glycol-sodium sulphate two-phase systems in the presence of iodide and thiocyanate ions. Anal Chim Acta 427:293–300
21. Cai C, Fajar ATN, Hanada T, Wakabayashi R, Goto M (2023) Amino acid leaching of critical metals from spent lithium-ion batteries followed by selective recovery of cobalt using aqueous biphasic system. ACS Omega 8:3198–3206
22. Shirayama S, Uda T (2015) Recovery of cobalt ion into polyethyleneglycol (PEG) gel phase as thiocyanato complex. Mater Trans 4:610–616
23. Ferreira GMD, Hespanhol MC, da Silva LHM, Patrício PR, da Cunha RC (2021) Elucidating mechanism for the extraction of cobalt, nickel, and iron in polymer-electrolyte aqueous biphasic systems with both phases rich in electrolyte. J Mol Liq 327:114803
24. Patrício PR, Mesquita MC, da Silva LHM, da Silva MCH (2011) Application of aqueous two-phase systems for the development of a new method of cobalt(II), iron(III) and nickel(II) extraction: a green chemistry approach. J Hazard Mater 193:311–318

25. Ishiguro S, Sotobayashi T, Satoh K, Saito K (1988) Calorimetric study of thiocyanato complexes of the cobalt(II) ion in micellar solutions of a nonionic surfactant. Inorg Chem 27:1152–1155
26. Umebayashi Y, Ishiguro, (1996) Spectrophotometric study of thiocyanato complexation of cobalt(II) and nickel(II) ions in micellar solutions of a nonionic surfactant Triton X-100. J Solution Chem 25:731–746
27. Shin M, Umebayashi Y, Ishiguro S (1997) Distribution thermodynamics of metal complexes in micelles of nonionic surfactants. Anal Sci 13:115–118
28. Umebayashi Y, Shin M, Kanzaki R, Ishiguro S (2001) Thermodynamics of $[Co(NCS)_4]^{2-}$ at poly(ethylene oxide) and octylphenyl moieties in micelles of nonionic surfactants. J Colloid Interf Sci 237:167–173
29. Shibukawa M, Hirasawa T, Saito S (2024) Structure of polyethylene glycol/water mixtures as a separation medium of aqueous two-phase extraction systems studied by equilibrium analysis of thiocyanato complexation of Cobalt(II). Polymer 293:126668
30. Silber H, Murguia MA (1985) Spectrophotometric investigations of the octahedral-tetrahedral equilibria in cobalt (II)-thiocyanate systems in aqueous methanol. Inorg Chem 24:3794–3802
31. Tribalat S, Zeller C (1962) Thiocyanate complexes of cobalt in water and in methyl isobutyl ketone. Bull Soc Chim France 11:2041–2047
32. Hammouda B, Ho DL, Kline S (2004) Insight into clustering in poly(ethylene oxide) solutions. Macromolecules 37:6932–6937
33. Cao N, Zhao Y, Chen H, Huang J, Yu M, Bao Y, Wang D, Cui S (2022) Poly(ethylene glycol) becomes a supra-polyelectrolyte by capturing hydronium ions in water. Macromolecules 55:4656–4664
34. Polik WF, Burchard W (1983) Static light scattering from aqueous poly(ethylene oxide) solutions in the temperature range 20–90°C. Macromolecules 16:978–982
35. Duval M (2000) Monitoring of cluster formation and elimination in PEO solutions. Macromolecules 33:7862–7867
36. Hammouda B, Ho D, Kline S (2002) SANS from poly(ethylene oxide)/water systems. Macromolecules 35:8578–8585
37. Polverari M, van de Ven TGM (1996) Dilute aqueous poly(ethylene oxide) solutions: clusters and single molecules in thermodynamic equilibrium. J Phys Chem 100:13696–13700
38. Gekko K (1983) Hydration-structure-function relationship of polysaccharides and proteins. In: Tanaka N, Ohtaki H, Tamamushi R (eds) Ions and molecules in solution. Elsevier, Amsterdam, pp 339–358
39. Gupta V, Nath S, Chand S (2002) Role of water structure on phase separation in polyelectrolyte-polyethyleneglycol based aqueous two-phase systems. Polymer 43:3387–3390
40. Ayranci E, Sahin M (2008) Interactions of polyethylene glycols with water studied by measurements of density and sound velocity. J Chem Themodyn 40:1200–1207
41. Derkaoui N, Said S, Grohens Y, Olier R, Privat M (2007) PEG400 novel phase description in water. J Colloid Interf Sci. 305:330–338
42. Sun P, Huang K, Lin J, Liu H (2018) Role of hydrophobic interaction in driving the partitioning of metal ions in a PEG-based aqueous two-phase system. Ind Eng Chem Res 57:11390–11398
43. Tasaki K (1999) Poly(oxyethylene)-cation interactions in aqueous solution: a molecular dynamics study. Comput Theor Polym Sci 9:271–284
44. Okada T (1990) Efficient evaluation of poly(oxyethylene) complex formation with alkali-metal cations. Macromolecules 23:4216–4219
45. Ries ME, Brereton MG, Cruickshank JM, Klein PG, Ward IM (1995) NMR study of poly(ethylene oxide) complexes with $LiCF_3SO_3$. Macromolecules 28:3282–3289
46. Marcus Y (2009) Effect of ions on the structure of water: structure making and breaking. Chem Rev 109:1346–1370
47. Edward JT, Farrell PG, Shahidi F (1977) Partial molar volumes of organic compounds in water. Part 1.-Ethers, ketones, esters and alcohols. J Chem Soc Faraday Trans 1 73:705–714
48. Berton P, Kelley SP, Bridges NJ, Klingshirn MA, Huddleston JG, Willauer HD, Baldwin JW, Moody ML, Rogers RD (2019) Water in solutions of chaotropic and kosmotropic salts: a differential scanning calorimetry investigation. J Chem Eng Data 64:4781–4792

49. Oesterhelt F, Rief M, Gaub HE (1999) Single molecule force spectroscopy by AFM indicates helical structure of poly(ethylene-glycol) in water. New J Phys 1:6
50. Shephard JJ, Bremer PJ, McQuillan AJ (2009) Structure and conformation in mixtures of methyl-terminated poly(ethylene oxide) and water. Principal component analysis and band fitting of infrared absorptions. J Phys Chem B 113:14229–14238
51. Tominaga T, Hishida M, Murakami D, Fujii Y, Tanaka M, Seto H (2022) Experimental evidence of slow mode water in the vicinity of poly(ethylene oxide) at physiological temperature. J Phys Chem B 126:1758–1767
52. Shikata T, Okuzono M, Sugimoto N (2013) Temperature-dependent hydration/dehydration behavior of poly(ethylene oxide)s in aqueous solution. Macromolecules 46:1956–1961
53. Smith GD, Bedrov D (2002) A molecular dynamics simulation study of the influence of hydrogen-bonding and polar interactions on hydration and conformations of a poly (ethylene oxide) oligomer in dilute aqueous solution. Macromolecules 35:5712–5719
54. Dahal UR, Dormidontova EE (2017) The dynamics of solvation dictates the conformation of polyethylene oxide in aqueous, isobutyric acid and binary solutions. Phys Chem Chem Phys 19:9823–9832
55. Dormidontova EE (2002) Role of competitive PEO-water and water-water hydrogen bonding in aqueous solution PEO behavior. Macromolecules 35:987–1001
56. Smith GD, Bedrov D, Borodin O (2000) Molecular dynamics simulation study of hydrogen bonding in aqueous poly(ethylene oxide) solutions. Phys Rev Lett 85:5583

Chapter 3
Effects of the Background Phase-Forming Electrolytes on the Partition of Ionic Solutes in PEG/Salt ATPS

Abstract The partitioning of nonionic compounds in the PEG/salt ATPS is predominantly determined by the difference in the concentration of PEG between the upper and lower phases, Δw_{PEG}, regardless of the type and concentration of the phase-forming salt. On the contrary, the distribution of ionic solutes depends on the type of the phase-forming salt even though the values of Δw_{PEG} are identical. The inherent effect of the background phase-forming cations and anions on the distribution of ionic solutes in polymer/salt ATPS must be evaluated under the conditions where the effect of the polymer concentration can be regarded as being constant. This chapter discusses the effects of counterion and co-ion of the phase-forming electrolyte on the distribution of ionic solutes based on the experimental results obtained for the PEG/salt ATPSs formed by PEG and Na_2SO_4, Na_2HPO_4, or K_2HPO_4. The abilities of K^+ and SO_4^{2-} to salt out the polymer are lower than those of Na^+ and HPO_4^{2-}, respectively, but the former ions are more effective than the latter ions in extracting oppositely charged ions into the PEG-rich phase when compared under constant Δw_{PEG}. On the other hand, Na^+ and HPO_4^{2-} are more effective than K^+ and SO_4^{2-} for the extraction of the ions which have the same charge sign as theirs. It will be shown that these effects of the background phase-forming ions on the distribution coefficient of an ionic solute can be interpreted by the stoichiometric ion partition model.

Keywords Aqueous polyethylene glycol/salt two-phase system · Liquid–liquid extraction · Counterion effect · Co-ion effect · Ionic solutes · Ion partition model

3.1 Introduction

PEG/salt ATPSs are formed when aqueous PEG solutions are salted out by salt solutions, producing two phases. Various salts can be used to form PEG/salt ATPS. It has been shown that the contributions of anions of the salts dominate in the two-phase formation compared to cations and their ability to form ATPS with PEG usually follows the Hofmeister series and can be related to the Gibbs free energy of hydration

M. Shibukawa, *Polymer-Based Aqueous-Two Phase Systems*,
https://doi.org/10.1007/978-981-96-2888-9_3

of ions, ΔG_{hyd} [1–13]; the more negative the ΔG_{hyd} of an ion, the greater its salting-out effect for PEG resulting in the lower concentrations of PEG and salt required to form ATPS. However, the partitioning of nonionic compounds in the PEG-based ATPS is predominantly determined by the difference in the volume fractions of the EO phase and the OH phase between the upper and lower phases regardless of the type and concentration of the salt as described in Chap. 2.

On the other hand, it has been shown that the distribution of ionic solutes in polymer/salt ATPS depends on the type of the phase-forming salt [13–20]. Rogers et al. [15] measured the distribution coefficients of pertechnetate ion, D_{Tc}, in ATPSs formed by mixing PEG and alkali metal or ammonium salts of SeO_4^{2-}, SO_3^{2-}, SO_4^{2-}, or CO_3^{2-}. They showed a linear relationship between $\log D_{Tc}$ and the total ΔG_{hyd} of the salt solution calculated by summing the molar ionic concentration times the ΔG_{hyd} of the individual ions and concluded that the D_{Tc} follows from the relative salting-out ability of the phase-forming salts. Bulgariu and Bulgariu investigated the extraction behavior of Hg(II), Cd(II), and Zn(II) in the ATPSs formed by PEG1500 and sulfate salt of Na^+ or NH_4^+ using iodide ion as the extracting agent and indicated that the PEG/$(NH_4)_2SO_4$ ATPS is more efficient than that containing Na_2SO_4 in the extraction of the metal ions [21]. They attributed the difference observed in the extraction efficiency of the metal ions due to the nature of the phase-forming salt to the differences of the salting-out effect of these salts on PEG, in addition to the differences in the concentrations of the salts used for preparation of the ATPSs.

However, the change in the distribution of ionic solutes in the PEG/salt ATPS with the nature of the phase-forming ions observed in most of the studies is primarily attributed to the change in the phase diagram or the difference in the PEG concentration between the upper and lower phases, Δw_{PEG}. To evaluate the intrinsic effect of the phase-forming ions as the background ions on the distribution of the target ions, the distribution coefficients should be measured under conditions where the Δw_{PEG} is constant [22]. In this chapter, the distribution behavior of inorganic cations and anions is discussed as a function of Δw_{PEG} in PEG/Na_2SO_4, K_2SO_4, and Na_2HPO_4 ATPSs on the basis of a model that we presented for the partition of ionic solutes [22, 23]. It will be shown that phase-forming ions exhibit intrinsic effects on the distribution of ionic solutes as the background ions and the effects can be interpreted by the stoichiometric ion partition model.

3.2 Effects of the Phase-Forming Salt on the Phase Diagram of PEG/Salt ATPS

Figure 3.1 shows phase diagrams for the ATPS containing Na_2SO_4, Na_2HPO_4, or K_2HPO_4 as a phase-forming salt and PEG4000 at 25 °C. The PEG concentrations are represented by weight percentage, whereas the salt concentrations are given in molality. The positions of the binodal lines shown in Fig. 3.1 demonstrate the effectiveness of cations and anions of the salts in inducing the formation of ATPS with PEG; $Na^+ > K^+$ for cations and $HPO_4^{2-} > SO_4^{2-}$ for anions. This order

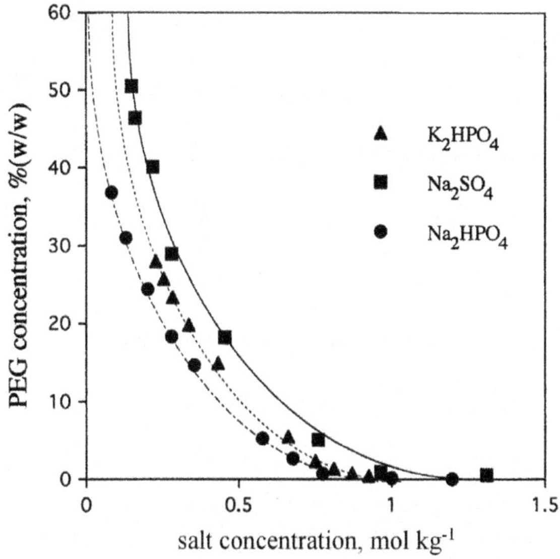

Fig. 3.1 Phase diagrams tor aqueous PEG/Na_2SO_4, Na_2HPO_4, and K_2HPO_4 two-phase systems. The weight percent of PEG4000 in each phase is plotted as the ordinate and the molality of the salt as the abscissa. Temperature: 25 °C (reproduced from Ref. [22] by permission of Springer Nature)

indicates the degree of incompatibility between the ion and PEG or the magnitude of the salting-out effect of the ion. As described above, the more negative the ΔG_{hyd} of the phase-forming ion, the binodal curve is shifted to the left corresponding to an increase in the area representing the two-phase area. The ΔG_{hyd} values for Na^+ and K^+ are −365 and −295 kJ mol^{-1}, respectively [24]. The ΔG_{hyd} value for $HPO_4{}^{2-}$ ion is not available but the smaller standard partial entropy of aqueous $HPO_4{}^{2-}$ ion (10.9 J K^{-1} mol^{-1}) compared to that of aqueous $SO_4{}^{2-}$ (64.5 J K^{-1} mol^{-1}) [25] suggests that $HPO_4{}^{2-}$ is more strongly hydrated than $SO_4{}^{2-}$. Experiments with aqueous solutions containing PEG of different molecular weights have also shown that $HPO_4{}^{2-}$ is more effective in salting PEG than $SO_4{}^{2-}$ [26, 27].

3.3 Intrinsic Effect of the Background Phase-Forming Ions on the Distribution of Ionic Solutes in PEG/Salt ATPS

The distribution of an ionic solute is represented by the distribution coefficient, *D*, defined as

$$D = \frac{C_{\mathrm{U}}}{C_{\mathrm{L}}} \tag{3.1}$$

where C_{U} and C_{L} are the concentrations of the ion in the upper phase (PEG-rich phase) and the lower phase (salt-rich phase), respectively. As described in the preceding chapter, the ln *D* values of solutes (both ionic and nonionic) are correlated to Δw_{PEG} by Eq. (3.2) [8, 10, 22]:

$$\ln D = a\Delta w_{PEG} \tag{3.2}$$

where a is a proportionality constant. Figures 3.2, 3.3 and 3.4 show the plots of $\ln D$ for IO_3^-, Br^-, NO_3^-, I^-, and SCN^- against Δw_{PEG} in the PEG/Na_2SO_4, Na_2HPO_4, and K_2HPO_4 ATPSs, respectively. IO_3^- is a kosmotropic ion [28–31] and is distributed in the salt-rich phase, while the other anions are chaotropic [24, 25, 28] and favor the PEG-rich phase. The order of the D values for these anions, $IO_3^- < Br^- \approx NO_3^- < I^- < SCN^-$, approximately corresponds to that of the ΔG_{hyd} values (the ΔG_{hyd} values for IO_3^-, Br^-, NO_3^-, I^-, and SCN^- are −400, −315, −300, −275, and −280 kJ mol^{-1}, respectively [24]). As described in Chap. 2, the EO phase in PEG aggregates plays a predominant role in the distribution of solute compounds in PEG/salt ATPS and chaotropic anions tend to be distributed into the PEG-rich phase. Therefore, the difference in the distribution coefficients of the inorganic anions observed can be interpreted in terms of the hydration of the anions.

The plots of $\ln D$ for Li^+ and Mg^{2+} versus Δw_{PEG} in the PEG/Na_2SO_4, Na_2HPO_4, and K_2HPO_4 systems are shown in Figs. 3.5, 3.6 and 3.7, respectively. Li^+ and Mg^{2+} are both kosmotropic ions [24, 25, 28] and are distributed in the salt-rich phases in all the ATPSs.

The results shown in Figs. 3.2, 3.3, 3.4, 3.5, 3.6 and 3.7 indicate that the distribution of inorganic ions in the PEG/salt ATPSs is mainly governed by Δw_{PEG} in a similar manner to nonionic solutes. However, it can be seen from these figures that the slope of the plot of $\ln D$ versus Δw_{PEG} depends on the type of the phase-forming cation and anion. The slopes of the $\ln D$ versus Δw_{PEG} plots for the anions are larger in the ATPS containing HPO_4^{2-} and K^+ than in the system containing SO_4^{2-} and Na^+, respectively, while the plots for the cations show the reverse tendency. In addition, the order of the D values for Li^+ and Mg^{2+} varies with the type of the phase-forming salt.

Shibukawa and Ohta [23] presented a stoichiometric model for the partition of ionic solutes in a liquid–liquid two-phase system derived on the basis of the charge balance, and clarified the effect of a background electrolyte on the retention of ionic solutes in liquid–liquid partition chromatography. According to the model, the distribution coefficient of the analyte or target ion A^{p+}, D_A^{YX}, in an ATPS containing background phase-forming cation Y^{m+} and anion X^{n-} is given by

$$\ln D_A^{YX} = \frac{1}{n}\ln K_{AX} - \frac{p}{m+n}\ln K_{YX} + k_A^{YX} \tag{3.3}$$

where k_A^{YX} is a term comprising the activity coefficients of A^{p+}, Y^{m+}, and X^{n-}, and K_{AX} and K_{YX} are the equilibrium constants for the following equilibria between the lower salt-rich phase (L) and the upper PEG-rich phase (U):

$$nA^{p+}(L) + pX^{n-}(L) \rightleftarrows nA^{p+}(U) + pX^{n-}(U) \tag{3.4}$$

$$nY^{m+}(L) + mX^{n-}(L) \rightleftarrows nY^{m+}(U) + mX^{n-}(U) \tag{3.5}$$

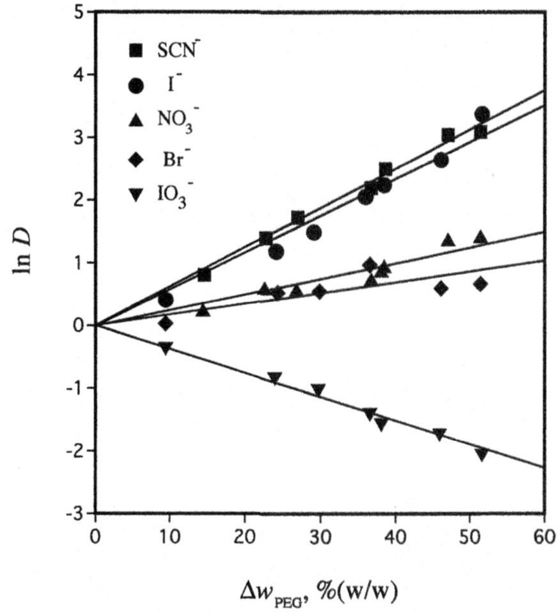

Fig. 3.2 ln *D* values of inorganic anions in PEG4000/Na_2SO_4 ATPS as a function of Δw_{PEG} (reproduced from Ref. [22] by permission of Springer Nature)

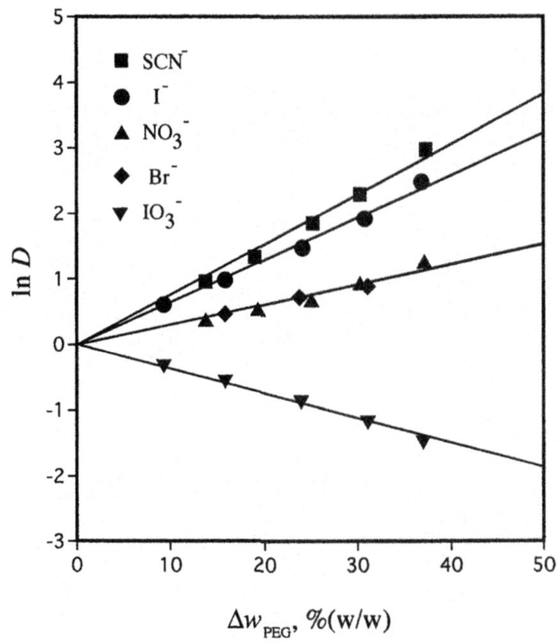

Fig. 3.3 ln *D* values of inorganic anions in PEG4000/Na_2HPO_4 ATPS as a function of Δw_{PEG} (reproduced from Ref. [22] by permission of Springer Nature)

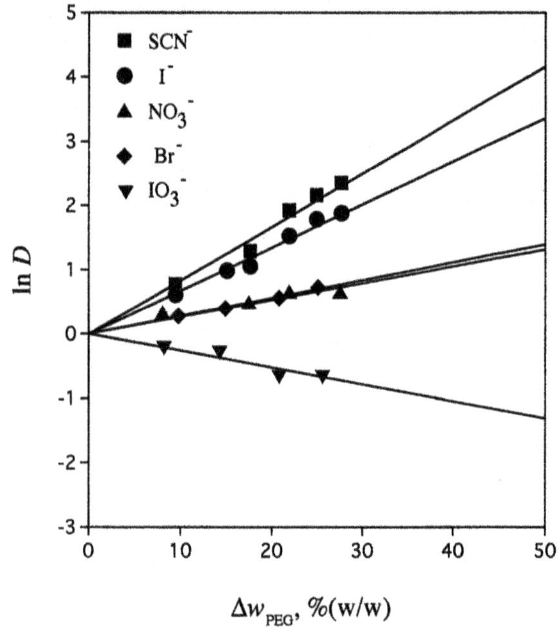

Fig. 3.4 ln D values of inorganic anions in PEG4000/K_2HPO_4 ATPS as a function of Δw_{PEG} (reproduced from Ref. [22] by permission of Springer Nature)

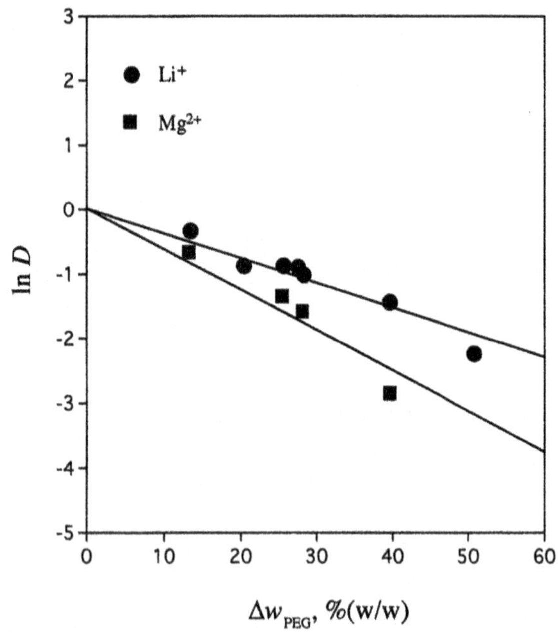

Fig. 3.5 ln D values of Li^+ and Mg^{2+} in PEG4000/Na_2SO_4 ATPS as a function of Δw_{PEG} (reproduced from Ref. [22] by permission of Springer Nature)

Fig. 3.6 ln D values of Li^+ and Mg^{2+} in PEG4000/Na_2HPO_4 ATPS as a function of Δw_{PEG} (reproduced from Ref. [22] by permission of Springer Nature)

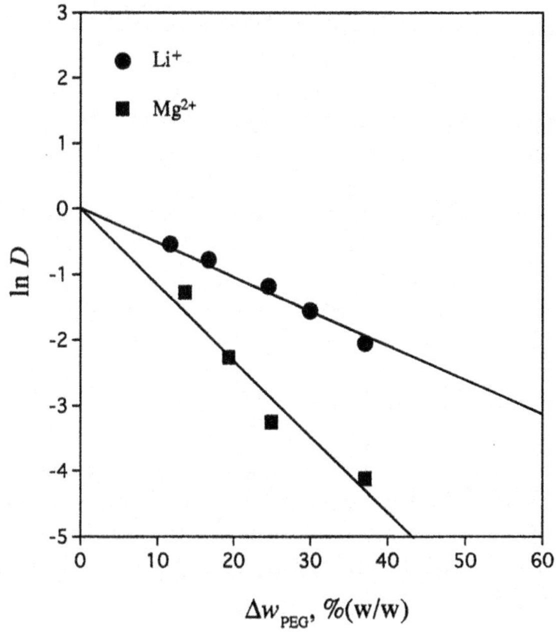

Fig. 3.7 ln D values of Li^+ and Mg^{2+} in PEG4000/K_2HPO_4 ATPS as a function of Δw_{PEG} (reproduced from Ref. [22] by permission of Springer Nature)

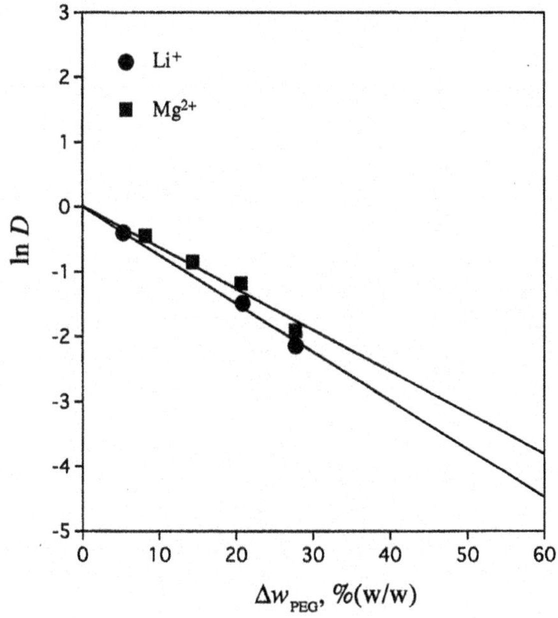

Equation (3.3) indicates that the larger the K_{AX} value (or the distribution coefficient of the counterion) and the smaller the K_{YX} value (or the distribution coefficient of the co-ion), the larger the distribution coefficient of the analyte ion, A^{p+}.

Equation (3.3) holds for the systems where the analyte ions do not form any complexes and ion pairs with counterions. Although the concentrations of the phase-forming salts are so high in usual PEG/salt ATPSs, the association reactions of the cations and anions with the counterions in the systems shown in Figs. 3.2, 3.3, 3.4, 3.5, 3.6 and 3.7 can be assumed to be negligible.

The effects of the phase-forming ions on the distribution of the analyte ions can be evaluated by the slopes of the $\ln D$ versus Δw_{PEG} plots, which represent the $\ln D$ values at Δw_{PEG} of 1%, $\ln D$ (1%). The $\ln D$ (1%) values for the analyte cations and anions in the three PEG/salt systems are tabulated in Table 3.1. $\ln D$ versus Δw_{PEG} plots for Na_2SO_4, Na_2HPO_4, and K_2HPO_4 in their respective ATPSs are shown in Fig. 3.8. All the plots give straight lines in the Δw_{PEG} range lower than 20%, although the data for the PEG/Na_2HPO_4 system form a convex curve in the higher Δw_{PEG} range. The D values for Na_2SO_4 in the PEG/Na_2SO_4 system, $D_{Na_2SO_4}^{Na_2SO_4}$, for example, represent the distribution coefficients of Na^+ and SO_4^{2-} in the system, since the D values for both ions are the same as each other in this ATPS ($D_{Na_2SO_4}^{Na_2SO_4} = D_{Na}^{Na_2SO_4} = D_{SO_4}^{Na_2SO_4}$). $K_{Na_2SO_4}$ is thus calculated as

$$\begin{aligned} K_{Na_2SO_4} &= \frac{C_{U,Na}^2 C_{U,SO_4}}{C_{L,Na}^2 C_{L,SO_4}} = (D_{Na}^{Na_2SO_4})^2 D_{SO_4}^{Na_2SO_4} = (D_{Na}^{Na_2SO_4})^3 \\ &= (D_{SO_4}^{Na_2SO_4})^3 = (D_{Na_2SO_4}^{Na_2SO_4})^3 \end{aligned} \tag{3.6}$$

$K_{Na_2HPO_4}$ and $K_{K_2HPO_4}$ can also be obtained in a similar manner. Figure 3.8 shows that the order of magnitude of the K values for the phase-forming salts in their ATPSs is $Na_2HPO_4 < K_2HPO_4 < Na_2SO_4$. This indicates that $K_{Na_2HPO_4} < K_{K_2HPO_4}$ and $K_{Na_2HPO_4} < K_{Na_2SO_4}$ or the distribution coefficients of K^+ and SO_4^{2-} are larger than those of Na^+ and HPO_4^{2-}, respectively. It can be seen from Table 3.1 that the distribution coefficients of the anions are larger in the ATPS formed by K^+ and HPO_4^{2-} than in the system containing Na^+ and SO_4^{2-}, respectively. On the contrary, the distribution coefficients of Li^+ are larger in the ATPS formed by Na^+ and SO_4^{2-} than those in the system formed by K^+ and HPO_4^{2-}, respectively. The distribution behavior of Mg^{2+} differs from that predicted from Eq. (3.3) in some cases, possibly because the activity coefficient of a polyvalent ion may not be regarded as constant independent of the type of the background electrolyte.

These results indicate that the dependence of the distribution of ionic solutes in PEG/salt ATPSs on the type of background phase-forming salt can be interpreted with Eq. (3.3) derived from the ion partition model. The distribution coefficient of an ion in the PEG/salt ATPS is larger when the background counterion of the phase-forming salt has the larger D value and the co-ion has the smaller value when compared under a constant Δw_{PEG}.

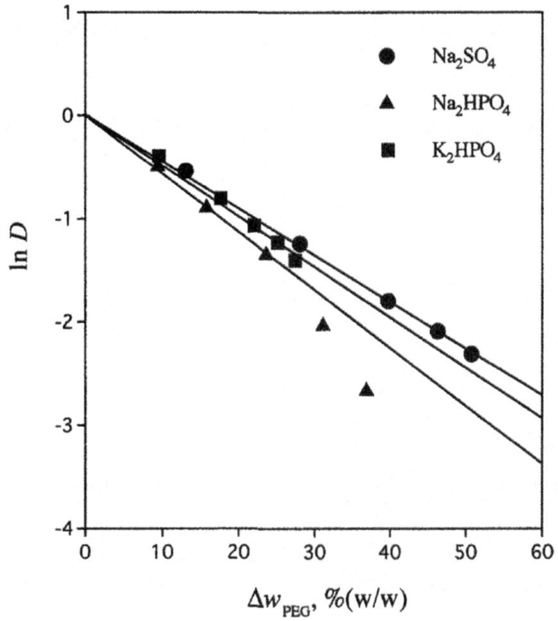

Fig. 3.8 ln *D* values of the phase-forming salts in PEG4000/Na_2SO_4, Na_2HPO_4, and K_2HPO_4 ATPSs as a function of Δw_{PEG} (reproduced from Ref. [22] by permission of Springer Nature)

3.4 Conclusions

The effectiveness of salts to form ATPS with PEG depends on their constituent cations and anions; the stronger is the hydration or the more negative the ΔG_{hyd} of the ion, the lower is the concentration required to form the two-phase system. Namely, Na^+ and HPO_4^{2-} are more effective than K^+ and SO_4^{2-} for salting out PEG, respectively. However, the latter ions are more effective for the extraction of oppositely charged analyte ions to the PEG-rich phase than the former ions when they were compared with each other under a constant Δw_{PEG}. On the contrary, Na^+ and HPO_4^{2-} are more effective than K^+ and SO_4^{2-} for the extraction of the ions which have the same charge sign as theirs. In most of the previous studies, these phase-forming electrolyte effects have been attributed to the different ion–ion interactions between the analyte ions and the phase-forming cations and anions and/or specific interactions between the analyte ions and the pseudopolycations formed by the metal ions and ethylene oxide chains of PEG molecules. However, the counterion and co-ion effects of the background phase-forming ions on the distribution of ionic solutes in the PEG/salt ATPS are rationally interpreted by the ion partition model regarding the distribution mechanism of ions in the presence of excess amounts of coexisting salts. The inherent effect of phase-forming salts on the partitioning of ionic solutes in ATPS should be evaluated under the conditions where the effect of the polymer concentrations can be regarded as being constant.

Table 3.1 ln *D* (1%) values for inorganic ions in PEG/salt ATPSs (reproduced from Ref. [22] by permission of Springer Nature)

Analyte ion	Phase-forming salt		
	Na_2SO_4	Na_2HPO_4	K_2HPO_4
IO_3^-	−0.038	−0.035	−0.026
Br^-	0.017	0.027	0.029
NO_3^-	0.025	0.027	0.028
I^-	0.059	0.064	0.067
SCN^-	0.063	0.073	0.083
Li^+	−0.038	−0.050	−0.074
Mg^{2+}	−0.063	−0.116	−0.066

References

1. Bailey Jr FE, Callard RW (1959) Some properties of poly(ethylene oxide) in aqueous solution. J Appl Polym Sci 1:56–62
2. Boucher EA, Hines PM (1976) Effects of inorganic salts on the properties of aqueous poly(ethylene oxide) solutions. J Polym Sci Polym Phys Ed 14:2241–2251
3. Ataman M, Boucher EA (1982) Properties of aqueous salt solutions of poly(ethylene oxide). J Polym Sci Polym Phys Ed 20:1585–1592
4. Florin E, Kjellander R, Eriksson JC (1984) Salt effects on the cloud point of the poly(ethylene oxide) + water system. J Chem Soc Faraday Trans 1(80):2889–2910
5. Ananthapadmanabhan KP, Goddard ED (1986) A correlation between clouding and aqueous biphase formation in polyethylene oxide/inorganic salt systems. J Colloid Interface Sci 113:294–296
6. Ananthapadmanabhan KP, Goddard ED (1987) Aqueous biphase formation in polyethylene oxide-inorganic salt systems. Langmuir 3:25–31
7. Huddleston JG, Griffin ST, Zhang J, Willauer HD, Rogers RD (1999) Metal ion separations in aqueous biphasic systems and using aqueous biphasic extraction chromatography. In: Bond AH, Diez ML, Rogers RD (eds) Metal-ion separation and preconcentration, ACS symposium series 716. American Chemical Society, Washington, pp 79–100
8. Willauer HD, Huddleston JG, Rogers RD (2002) Solute partitioning in aqueous biphasic systems composed of polyethylene glycol and salt: the partitioning of small neutral organic species. Ind Eng Chem Res 41:1892–1904
9. Willauer HD, Huddleston JG, Rogers RD (2002) Solvent properties of aqueous biphasic systems composed of polyethylene glycol and salt characterized by the free energy of transfer of a methylene group between the phases and by a linear solvation energy relationship. Ind Eng Chem Res 41:2591–2601
10. da Silva LHM, da Silva MCH, de Aquino RAN, Francisco KR, Cardoso MVC, Minim LA, Coimbra JSR (2006) Nitroprusside-PEO enthalpic interaction as a driving force for partitioning of the $[Fe(CN)_5NO]^{2-}$ anion in aqueous two-phase systems formed by poly(ethylene oxide) and sulfate salts. J Phys Chem B 110:23540–23546
11. Ferreira LA, Teixeira JA (2011) Salt effect on the aqueous two-phase system PEG 8000-sodium sulfate. J Chem Eng Data 56:133–137
12. Berggren K, Johansson HO, Tjerneld F (1995) Effects of salts and the surface hydrophobicity of proteins on partitioning in aqueous two-phase systems containing thermoseparating ethylene oxide-propylene oxide copolymers. J Chromatogr A 718:67–79
13. Rogers RD, Bond AH, Bauer CB, Zhang J, Griffin ST (1996) Metal ion separations in polyethylene glycol-based aqueous biphasic systems: correlation of partitioning behavior with available thermodynamic hydration data. J Chromatogr B 680:221–229

14. Johansson G (1985) Determination of ionic charge by liquid-liquid partition. J Chromatogr 322:425–432
15. Rogers RD, Zhang J (1996) Effects of increasing polymer hydrophobicity on distribution ratios of TcO_4^- in polyethylene/poly(propylene glycol)-based aqueous biphasic systems. J Chromatogr B 680:231–236
16. Patrício PR, Cunha RC, Vargas SJR, Coelho YC, da Silva LHM, da Silva MCH (2016) Chromium speciation using aqueous biphasic systems: development and mechanistic aspects. Sep Purif Technol 158:144–154
17. Paik SP, Sen K (2016) Species dependent iodine extractions in polymer based aqueous biphasic systems: emerging relations with aggregation number of polymeric micelles. J Mol Liq 223:1062–1066
18. Hey MJ, Jackson DP, Yan H (2005) The salting-out effect and phase separation in aqueous solutions of electrolytes and poly(ethylene glycol). Polymer 46:2567–2572
19. Huddleston JG, Willauer HD, Rogers RD (2003) Phase diagram data for several PEG + salt aqueous biphasic systems at 25 °C. J Chem Eng Data 48:1230–1236
20. Silvério SC, Madelra PP, Rodriguez O, Teixeira JA, Macedo EA (2008) $\Delta G(CH_2)$ in PEG-salt and Ucon-salt aqueous two-phase systems. J Chem Eng Data 53:1622–1625
21. Bulgariu L, Bulgariu D (2013) Selective extraction of Hg(II), Cd(II) and Zn(II) ions from aqueous media by a green chemistry procedure using aqueous two-phase systems. Sep Purif Technol 118:209–216
22. Shibukawa M, Matsuura K, Shinozuka Y, Mizuno S, Oguma K (2000) Effects of phase-forming cations and anions on the partition of ionic solutes in aqueous polyethylene glycol-inorganic salt two-phase systems. Anal Sci 16:1039–1044
23. Shibukawa M, Ohta N (1986) Model for partition chromatography of ionic solutes in the presence of eluent electrolytes. Chromatographia 22:261–267
24. Marcus Y (1991) Thermodynamics of solvation of ions Part 5. Gibbs free energy of hydration at 298.15 K. J Chem Soc Faraday Trans 87:2995–2999
25. Marcus Y (1985) Ion solvation. Wiley-Interscience, Chichester
26. Spivakov BY, Nifant'eva TI, Shikinev VM (1995) Metal extraction in two-phase water-poly(ethylene glycol)-salt systems. In: Rogers RD, Eiteman MA (eds) Aqueous biphasic separations: biomolecules to metal ions. Plenum Press, New York, pp 83–90
27. Eiteman MA (1995) Hydrophobic and charge effects in the partitioning of solutes in aqueous two-phase systems. In: Rogers RD, Eiteman MA (eds) Aqueous biphasic separations: biomolecules to metal ions. Plenum Press, New York, pp 31–48
28. Nightingale ER Jr (1966) On the specificity of electrolyte solvation. Viscosity and infrared characterization of ionic hydration. In: Conway BE, Barradas RG (eds) Chemical physics of ionic solutions. Wiley, New York, pp 87–100
29. Jenkins HDB, Marcus Y (1995) Viscosity B-coefficients of ions in solution. Chem Rev 95:2695–2724
30. Baer MD, Pham V, Fulton JL, Schenter GK, Balasubramanian M, Mundy CJ (2011) Is iodate a strongly hydrated cation? J Phys Chem Lett 2:2650–2654
31. dos Santos AP, Diehl A, Levin Y (2010) Surface tensions, surface potentials, and the Hofmeister series of electrolyte solutions. Langmuir 26:10778–10783

Chapter 4
Trace Analyses of Metal Ions in Environmental Samples Based on the Separation and Preconcentration by Extraction with ATPS

Abstract Liquid–liquid extraction with PEG-based ATPS is a promising strategy for the recovery and separation of metal ions from aqueous media replacing conventional organic-aqueous systems from an environmental point of view. Although ATPS extraction methods using an organic ligand that coordinates metal ions have been investigated since this technique is commonly used in conventional liquid–liquid extraction of metal ions, the use of this method for actual analytical or industrial separations is very limited because most of the organic extracting agents are hydrophobic and do not dissolve in water. On the other hand, extraction of metal ions as negatively charged complexes of halide and pseudohalide ions has been proven to be effective in PEG/salt ATPS. In the present chapter, the mechanism of the extraction of metal ions in PEG/Na_2SO_4 ATPS is discussed taking iodide and thiocyanate ions as the extracting agents based on the complexation of the metal ions not only with iodide or thiocyanate ion but also with the phase-forming sulfate ion. Simple and green extraction-spectrophotometric methods using PEG/Na_2SO_4 ATPS developed for on-site analyses of heavy metal ions in environmental samples (river water, soil, and sediment) will also be presented.

Keywords Aqueous two-phase system · Metal extraction · Extraction mechanism · Halide and pseudohalide ions · On-site environmental analysis

4.1 Introduction

PEG-based ATPS is a promising strategy for the enrichment and separation of metal ions in various samples replacing conventional organic-aqueous systems due to their environmentally benign advantages [1–33]. Much work has been carried out to investigate the distribution behavior of various metal species including lanthanides [1, 6, 23, 25] and actinides [1, 4, 15] in the ATPS. However, the poor extraction and separation efficiency of some metal ions have limited the widespread application of the ATPS to the analysis of trace metal ions in complex environmental and industrial samples.

M. Shibukawa, *Polymer-Based Aqueous-Two Phase Systems*,
https://doi.org/10.1007/978-981-96-2888-9_4

Generally, most metal ions stay in the salt-rich phases in PEG/salt ATPSs because of their strong hydrated structures. As described in Chap. 2, the pseudo-phase formed by the hydration of the ethylene oxide (EO) chains in the PEG molecules plays a predominant role in the partition of solute compounds into the PEG-rich phase in PEG/salt ATPS. It is thus necessary to modify the structures of target metal species to those which have high affinity to the EO phase. With $Co(NCS)_4{}^{2-}$, thiocyanate ion coordinates through the nitrogen atom, and accordingly the large sulfur end directs outward to make this complex hydrophobic [34, 35]. Therefore, metal complexes coordinated with hydrophobic ligands may be favored in PEG-rich phase. Some metal oxoanions such as $TcO_4{}^-$ and $HCrO_4{}^-$ and inorganic anions such as thiocyanate ion and iodide ion, known to be chaotropic ions [36], can also distribute into the polymer-rich phase without any extractant. These anionic chaotropic species may be located at a hydrophobic area of the EO pseudo-phase out of the PEG helix as counterions of the cations such as Na^+ and $NH_4{}^+$ coordinated with oxygen atoms of EO chains.

Extraction using an organic ligand which coordinates metal ions has also been studied as this technique is very commonly used in conventional organic-aqueous two-phase extractions. The PEG-rich phase has a hydrophobic domain in their structures and is able to dissolve some hardly water-soluble extractants and their metal complexes. However, the use of this method for actual analytical or industrial separations is very limited, although the distribution behavior of metal ions in ATPSs with organic extractants such as 1-(2-pyridylazo)-2-naphtol [29, 33], crown ether [15], and dimethylglyoxime [17] has been examined and some of them are shown to be applicable to metal analysis in industrial samples [17].

Metal ion extraction as negatively charged complexes of inorganic anions has been proven to be effective in the PEG/salt ATPS. Most of the soft metal ions can form the complexes with halide and pseudohalide ions and are extracted into the PEG-rich phase at high distribution coefficients. The conditions for selective extraction of the target metal ion by this technique can be optimized by the concentration of the extractant anion and with the use of suitable masking agents. In this chapter, the mechanism of the extraction of metal ions with inorganic anions in the PEG/salt ATPS is discussed taking iodide and thiocyanate ions as the extracting agents. Analytical methods for the determination of trace amounts of metal ions in environmental samples will also be described [24, 37].

Heavy metal ions are generally present in aquatic ecosystems and soil environments in trace amounts. However, most of them are highly toxic and have a tendency to bioaccumulate as a result of food chain [38, 39]. Accurate determination of heavy metal ions at trace levels in environmental samples has thus been required. Spectrometric methods such as atomic absorption spectrometry, atomic emission spectrometry, and inductively coupled plasma mass spectrometry are known to be sensitive analytical methods for trace metal ions. Although these methods are highly sensitive, they need expensive instrumentations and involve time-consuming procedures. Furthermore, these methods cannot be used for on-site measurements.

Nowadays, due to environmental concerns, fast on-site analyses of environmental samples are required [40]. UV-visible spectrophotometric method has an advantage that it is capable of performing fast on-site quantitative analyses with the inexpensive devises and simple procedures [41–43]. However, the conventional spectrophotometric methods lack selectivity and suffer from serious interference of foreign metal ions in environmental samples. As the PEG/salt ATPS has great potential for separating metal ions, interferences from the foreign metal ions are expected to be effectively removed by extraction with the ATPS. Simple and green extraction-spectrophotometric methods using the PEG/salt ATPS developed for on-site analyses of heavy metal ions will be presented in this chapter.

4.2 Distribution Mechanism of Metal Ions by Complexation with Inorganic Anions

Sodium sulfate exerts a strong salting-out effect for PEG and its high solubility in water allows the preparation of ATPSs with large Δw_{PEG}. This means that we can obtain an effective extraction system for analyte compounds using PEG and Na_2SO_4 as ATPS phase-forming components.

The PEG/Na_2SO_4 ATPSs used for the investigation of the distribution mechanism of metal ions were prepared by mixing 3.0 g of 50% (w/w) PEG4000 solution, 6.0 g of 20% (w/w) Na_2SO_4 solution, and 0.1 g of 1% (w/w) of H_2SO_4 in graduated glass centrifuge tubes with stoppers. Aliquots of a stock solution of a metal ion (100–250 μL) and 2 mol L^{-1} KI (0–250 μL) or 10 mol L^{-1} KSCN solution (0–500 μL) were then added to the ATPS prepared. Figure 4.1 shows the phase diagram of the PEG4000/Na_2SO_4 ATPS at 25 °C. The points U and L represent the composition of the upper phase and that of the lower phase in the extraction system used, respectively [16].

Figures 4.2 and 4.3 show the values of percent extraction for Cu(II), Zn(II), Fe(III), Co(II), Cd(II), Ni(II), Mn(II), and Li(I) plotted against the amount of SCN^- and I^- in the PEG4000/Na_2SO_4 ATPS, respectively [16]. The percent extraction of each metal ion is given by Eq. (4.1):

$$\%\text{Extraction} = \frac{n_{\text{M, PEG}}}{n_{\text{M, Total}}} \times 100 \tag{4.1}$$

where $n_{\text{M, PEG}}$ is the amount of a metal ion in the PEG-rich phase in mol and $n_{\text{M, Total}}$ is the total amount of the metal ion added to the system. In the ATPS with SCN^-, most transition metal ions were extracted into the PEG-rich phase, while in the I^- system a clear increase in extraction was observed only for Cd(II).

The enhancement of the extractability of the metal ions by the addition of SCN^- and I^- is attributed to the formation of metal complexes with chaotropic thiocyanate and iodide ions. The extraction of a metal ion, M^{m+}, by the formation of the complexes with an anion, X^-, can be described by the following equations:

$$M^{m+} + nX^- \rightleftarrows MX_n^{(n-m)-} \tag{4.2}$$

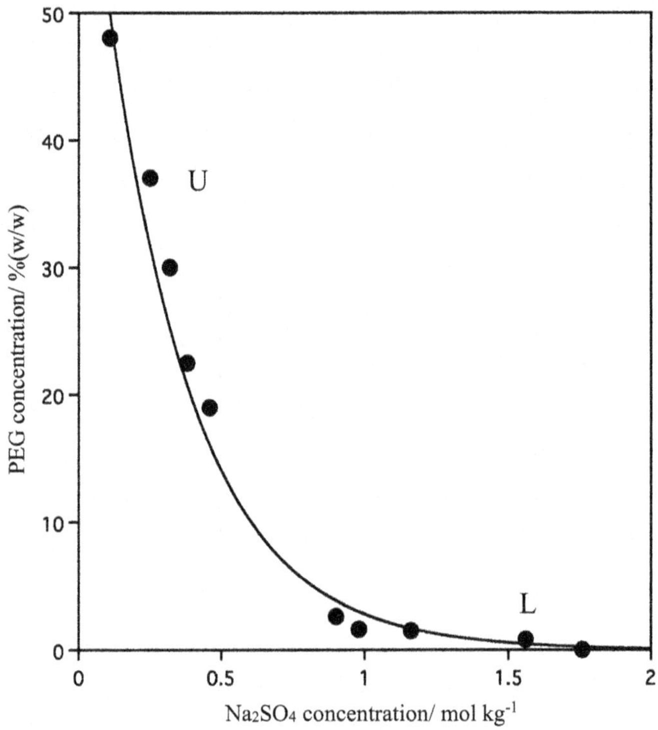

Fig. 4.1 Phase diagram for PEG/Na_2SO_4 ATPS. The weight percent of PEG4000 in each phase is plotted as the ordinate and the molality of the salt as the abscissa. The points U and L denote the upper and the lower phases in the present extraction system, respectively. Temperature: 25.0 ± 0.1 °C (reproduced from Ref. [16] by permission of Elsevier)

$$\beta_{MX_n} = \frac{[MX_n^{(n-m)-}]}{[M^{m+}][X^-]^n} \tag{4.3}$$

$$MX_n^{(n-m)-} \rightleftarrows (MX_n^{(n-m)-})_{PEG} \tag{4.4}$$

$$D_{MX_n} = \frac{[MX_n^{(n-m)-}]_{PEG}}{[MX_n^{(n-m)-}]} \tag{4.5}$$

where β_{MX_n} and D_{MX_n} are the overall formation constant and the distribution coefficient of the metal complex $MX_n{}^{(n-m)-}$ in the PEG/Na_2SO_4 ATPS, respectively. The subscript "PEG" denotes chemical species in the PEG-rich phase, while the species in the Na_2SO_4-rich phase is given without any subscript. The metal ion to be extracted as an anionic complex is distributed into the PEG-rich phase along with its counterion, sodium ion, while sodium ion is distributed with its counterion of the phase-forming salt, sulfate ion. The concentration of hydrogen ion in the present

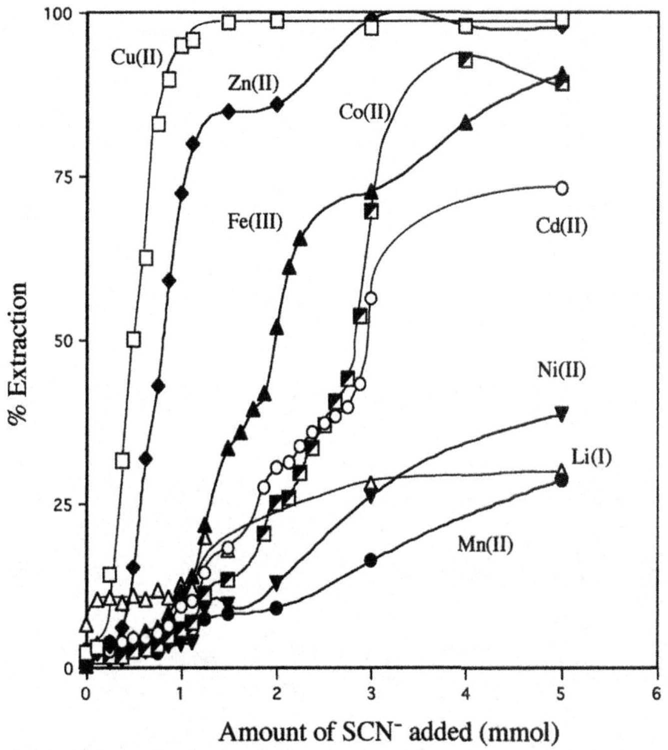

Fig. 4.2 Extraction of metal ions with thiocyanate ion in ATPS formed by mixing 3.0 g of 50% (w/w) PEG4000, 6.0 g of 20% (w/w) Na_2SO_4, 0.1 g of 1% (w/w) H_2SO_4, and 1 mL of water containing KSCN. The initial concentrations of metal ions in the extraction system were 10–25 ppm. Temperature: 25.0 ± 0.1°C (reproduced from Ref. [16] by permission of Elsevier)

ATPS is negligibly small compared to that of sodium ion. Therefore, as described in Chap. 3 (Eq. (3.3) [44]), D_{MX_n} is given by

$$\ln D_{MX_n} = \frac{1}{n-m} \ln K_{Na_{n-m}MX_n} - \frac{n-m}{3} \ln K_{Na_2SO_4} + k_{MX_n}^{Na_2SO_4} \tag{4.6}$$

where $K_{Na_{n-m}MX_n}$ and $K_{Na_2SO_4}$ are equilibrium constants given by the following equations:

$$K_{Na_{n-m}MX_n} = \frac{[Na^+]_{PEG}^{n-m}[MX_n^{(n-m)-}]_{PEG}}{[Na^+]^{n-m}[MX_n^{(n-m)-}]} \tag{4.7}$$

$$K_{Na_2SO_4} = \frac{[Na^+]_{PEG}^2[SO_4^{2-}]_{PEG}}{[Na^+]^2[SO_4^{2-}]} \tag{4.8}$$

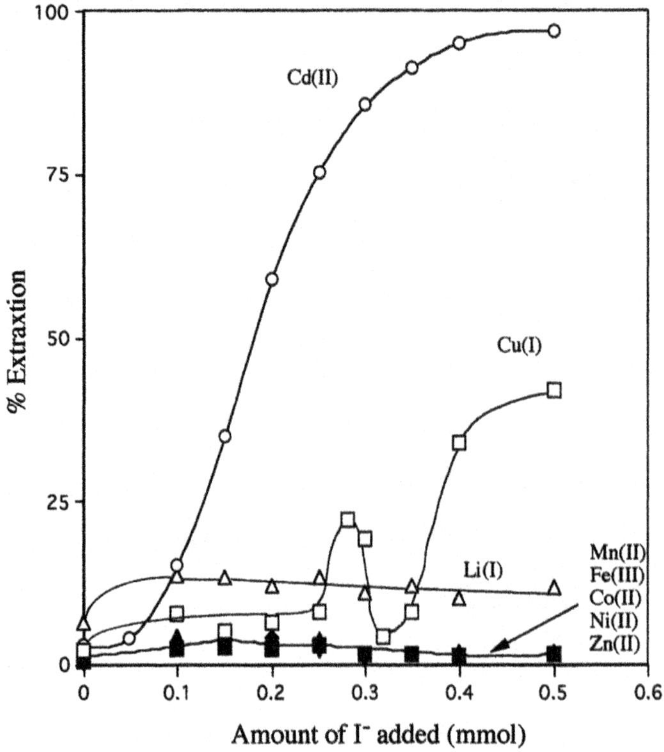

Fig. 4.3 Extraction of metal ions with iodide ion in the PEG/Na_2SO_4 ATPS. Composition of the extraction system except for KSCN and other experimental conditions are the same as in Fig. 4.2 (reproduced from Ref. [16] by permission of Elsevier)

The concentrations of sodium ion and sulfate ion in the upper and lower phases can be assumed to be constant since their concentrations are much larger than those of the other ionic species. It is thus reasonable to regard D_{MX_n} as constant.

Sulfate ion forms complexes with a number of metal ions so that the influence of the complexation reaction of metal ions with SO_4^{2-} on metal extractions is not negligible in PEG/Na_2SO_4 ATPS. The complexation reaction of a metal ion with SO_4^{2-} to form a complex can be represented as

$$M^{m+} + xSO_4^{2-} \rightleftarrows M(SO_4)_x^{(2x-m)-} \tag{4.9}$$

The formation constant of the metal sulfate complex is given by

$$\beta_{M(SO_4)_x} = \frac{[M(SO_4)_x^{(2x-m)-}]}{[M^{m+}][SO_4^{2-}]^x} \tag{4.10}$$

The distribution equilibrium and distribution coefficient of the metal sulfate complex are represented by Eqs. (4.11) and (4.12), respectively:

$$M(SO_4)_x^{(2x-m)-} \rightleftarrows (M(SO_4)_x^{(2x-m)-})_{PEG} \tag{4.11}$$

$$D_{M(SO_4)_x} = \frac{[M(SO_4)_x^{(2x-m)-}]_{PEG}}{[M(SO_4)_x^{(2x-m)-}]} \tag{4.12}$$

As a result, the distribution ratio of a metal ion, D_M, defined as the ratio of the metal concentration in the PEG-rich phase to that in the salt-rich phase can be represented as

$$D_M = \frac{\Sigma\left[MX_n^{(n-m)-}\right]_{PEG} + \Sigma\left[M(SO_4)_x^{(2x-m)-}\right]_{PEG}}{\left[M^{m+}\right] + \Sigma\left[MX_n^{(n-m)-}\right] + \Sigma\left[M(SO_4)_x^{(2x-m)-}\right]} \tag{4.13}$$

The concentration of $SO_4{}^{2-}$ in the salt-rich phase of the PEG/Na_2SO_4 ATPS used is 1.6 mol kg^{-1} as shown in Fig. 4.1, whereas those of SCN^- and I^- are less than 0.1 and 0.02 mol kg^{-1}, respectively. In addition, the first-order formation constant of the complex of each transition metal ion with $SO_4{}^{2-}$ is larger than or of the same order as that for the complex with SCN^- or I^- [45]. Therefore, it can be concluded that the transition metal ions are predominantly present in the form of sulfate complexes in the Na_2SO_4-rich phase. On the other hand, it is reasonable to assume that the concentration of the sulfate complex is negligibly small in the PEG-rich phase because the metal ions are scarcely extracted into the PEG-rich phase in the absence of SCN^- or I^-. Therefore, Eq. (4.13) can be rewritten as

$$D_M = \frac{\Sigma\left[MX_n^{(n-m)-}\right]_{PEG}}{\Sigma\left[M(SO_4)_x^{(2x-m)-}\right]} = \frac{\Sigma K_{ex,MX_n}\left[X^-\right]^n_{PEG}}{\Sigma\beta_{M(SO_4)_x}\left[SO_4^{2-}\right]^x} \tag{4.14}$$

where K_{ex,MX_n} is the extraction constant of $MX_n{}^{(n-m)-}$ given by

$$K_{ex,MX_n} = \frac{D_{MX_n}\beta_{MX_n}[X^-]}{[X^-]_{PEG}} = \frac{D_{MX_n}\beta_{MX_n}}{D_X} \tag{4.15}$$

Equation (4.14) indicates that the slope of a plot of $\log D_M$ against $\log[X^-]_{PEG}$ will give the composition of the metal complex extracted into the PEG-rich phase or the number of X^- ions per metal atom. The plots of $\log D$ versus $\log[SCN^-]_{PEG}$ are presented in Figs. 4.4 and 4.5. Straight lines with slopes around 4 are found for the plots of Cu(II), Fe(III), Zn(II), and Co(II) as shown in Fig. 4.4, indicating that four-coordinated thiocyanate complexes are formed for these metal ions. On the other hand, Fig. 4.5 shows that the plots for Cd(II), Ni(II), and Mn(II) give straight lines with the slope of 2, which indicates that the main extracted species of these metal ions in the ATPS containing SCN^- have a molar ratio M:SCN of 1:2. It has been known that hard metal ions such as Co(II) and Ni(II) tend to form *N*-bonded isothiocyanate complexes, whereas soft metal ions such as Cd(II) tend to form *S*-bonded thiocyanate complexes. However, the difference between these two structures is not apparent in these results.

Figure 4.6 shows the logD_M versus log$[I^-]_{PEG}$ plots for Li(I), Cu(II), and Cd(II). The slope of the plots for Cd(II) is 4, suggesting that Cd(II) is extracted as the complex CdI_4^{2-} in the ATPS containing I^-. On the other hand, the plots for Cu(II) show irregular changes over the iodide concentration range tested. This phenomenon may be due to the partial reduction of Cu(II) to Cu(I) by iodide ions.

The extractability of a metal ion in PEG/Na_2SO_4 ATPSs depends not only on the formation constant of the metal complex with the extraction reagent, i.e., SCN^- or I^-, but also on the formation constant of the sulfate complex. The formation constant of the complex of Fe(III) with SCN^- is the largest among the metal ions studied [45], but the distribution ratio of Fe(III) is smaller than those of Cu(II) and Zn(II). This is attributed to the formation of stable Fe(III) sulfate complexes [45] that prefer the salt-rich phase in the ATPS.

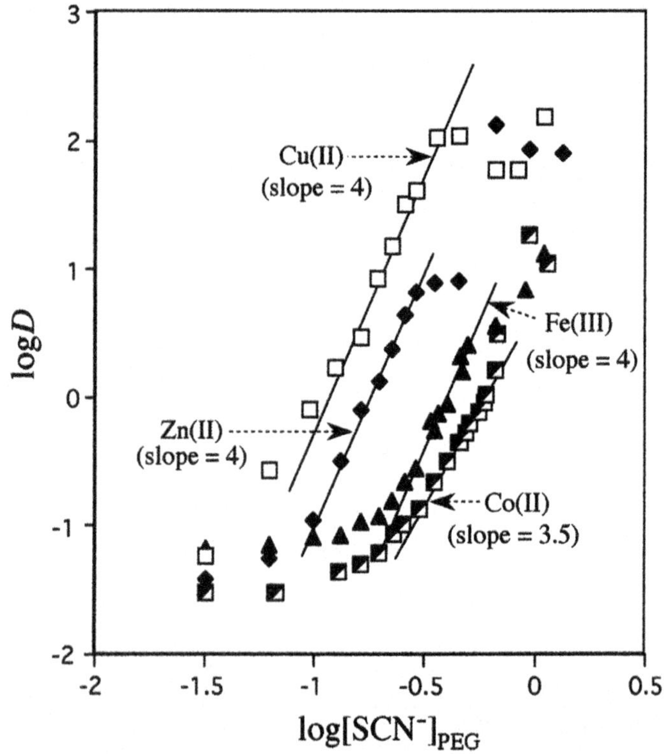

Fig. 4.4 Plots of logD_M against log$[SCN^-]_{PEG}$ for Fe(III), Co(II), Cu(II), and Zn(II). Composition of the extraction system and other experimental conditions are the same as in Fig. 4.2 (reproduced from Ref. [16] by permission of Elsevier)

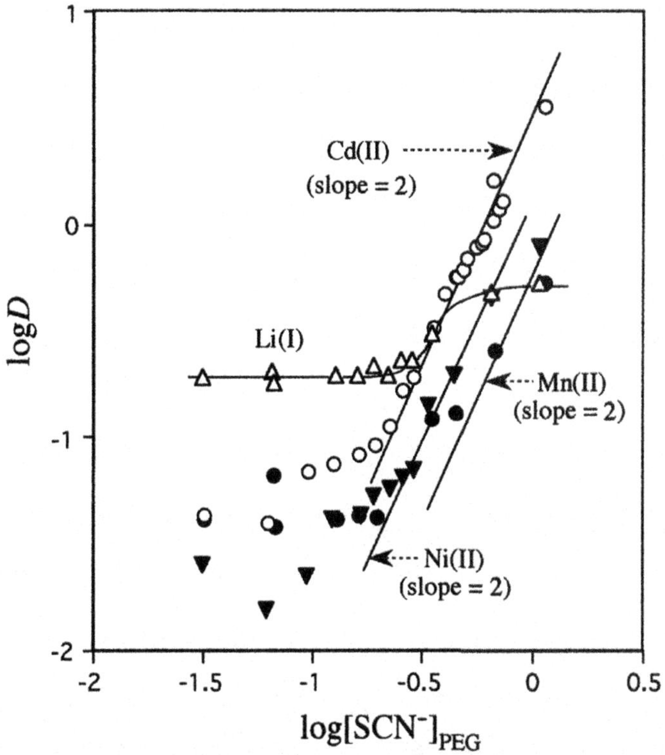

Fig. 4.5 Plots of logD_M against log[SCN^-]$_{PEG}$ for Li(I), Mn(II), Ni(II), and Cd(II). Composition of the extraction system and other experimental conditions are the same as in Fig. 4.2 (reproduced from Ref. [16] by permission of Elsevier)

Excessively high concentrations of extracting reagents added to the ATPS may alter the phase diagram of the system. Figures 4.4 and 4.5 show that discontinuities are observed in the log$D$ versus log[SCN^-]$_{PEG}$ plots in the region between −0.4 and −0.2 of log[SCN^-]$_{PEG}$. The distribution ratio of Li(I) also changes in this concentration range of SCN^- (Fig. 4.5). These discontinuities of the distribution ratios of the metal ions are attributed to changes of the phase diagram of the ATPS or the compositions of the two phases caused by the addition of excessive amount of KSCN. The variation of the volume ratio of the PEG-rich phase to the Na_2SO_4-rich phase, V_{PEG}/V_{salt}, with log[SCN^-]$_{PEG}$ is shown in Fig. 4.7. The V_{PEG}/V_{salt} value sharply increases at around log[SCN^-]$_{PEG}$ of −0.4, suggesting that a change of the phase compositions takes place when thiocyanate ion is added at [SCN^-]$_{PEG}$ > $10^{-0.4}$ mol L^{-1}. The increase in the volume of the PEG-rich phase results in the decrease in PEG concentration in the phase and thus Δw_{PEG} is also decreased. Thus, this volume change leads to a reduction of the extraction selectivity of the ATPS.

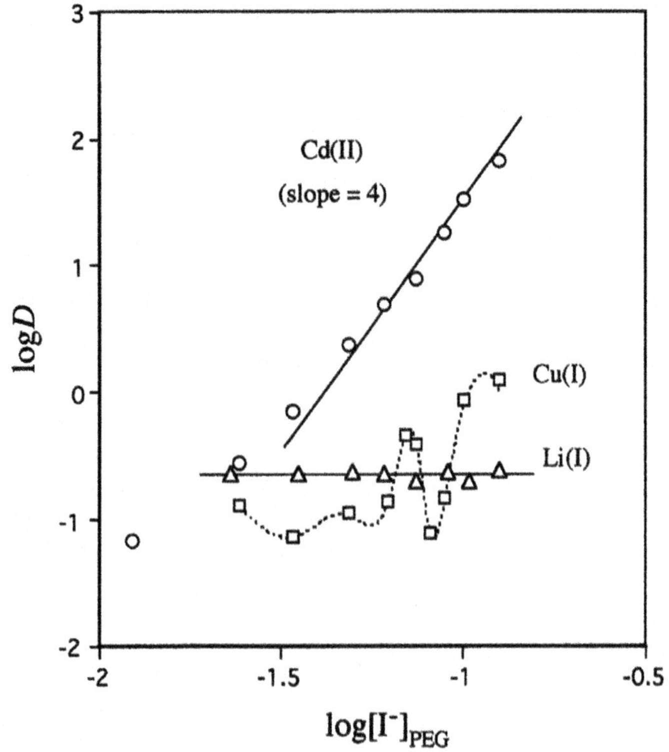

Fig. 4.6 Plots of log$D$ against log$[I^-]_{PEG}$ for Cd(II), Cu(I), and Li(I). Composition of the extraction system except for KI and other experimental conditions are the same as in Fig. 4.3 (reproduced from Ref. [16] by permission of Elsevier)

4.3 Spectrophotometric Determination of Trace Amounts of Cadmium in Soil and Sediment Samples

4.3.1 Extraction of Cd(II) with Iodide Ion in PEG6000/ Na_2SO_4 ATPS

Cadmium is a toxic metal and is well known to have adverse effects on the human body even in trace amounts [46–48]. Environmental pollution from Cd(II) is caused by anthropogenic and naturally occurring sources. Soils polluted with Cd(II) may cause serious hazards to the health of humans through intake of Cd(II)-contaminated crops grown in these soils. For this reason, accurate and precise determination of trace amounts of Cd(II) in soils is demanded [41, 49, 50].

UV-visible spectrophotometric method possesses distinct advantages over other frequently adopted methods such as atomic absorption spectrometry, atomic emission spectrometry, and inductively coupled plasma mass spectrometry with respect

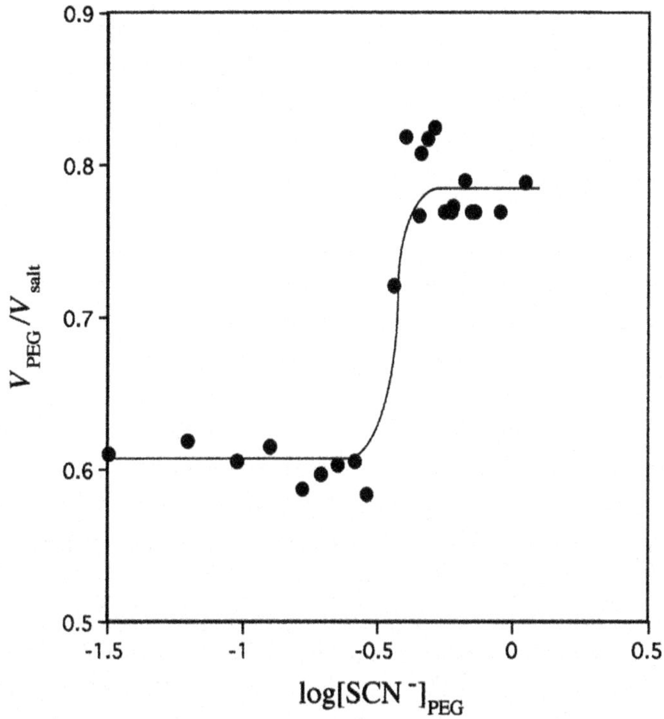

Fig. 4.7 Dependence of the ratio of the volume of the PEG-rich phase to that of the Na_2SO_4-rich phase on $\log[SCN^-]_{PEG}$ (reproduced from Ref. [16] by permission of Elsevier)

to simplicity, speed, and ease of operation. On the other hand, as shown in Sect. 4.2, Cd(II) is selectively extracted into the PEG-rich phase in PEG/Na_2SO_4 ATPS using iodide ion as an extractant (Fig. 4.3) [16]. It is thus expected that a simple spectrophotometric method capable of performing fast on-site quantitative analyses of trace amounts of Cd(II) in soils can be developed employing the PEG/Na_2SO_4 two-phase extraction for a pretreatment of the samples.

PEG6000 was adopted as a polymer component of the ATPS for the analytical method development, because the distribution ratio of Cd(II) obtained with PEG4000/Na_2SO_4 ATPS is not high enough for quantitative extraction of Cd(II). As the molecular weight of PEG increases, phase separation occurs at lower concentrations of PEG, resulting in larger Δw_{PEG} or ΔEO, which is expected to increase the D_M value of Cd(II) and decrease the D_M values of the other metal ions [1, 44]. Figure 4.8 shows the phase diagram for the PEG6000/Na_2SO_4 ATPS.

The dependence of the extraction of Cd(II) on the amount of iodide ion in the PEG6000/Na_2SO_4 ATPS is represented in Fig. 4.9. The ATPS was prepared by mixing 1 g of 50%(w/w) PEG6000 and 7 g of 20% (w/w) Na_2SO_4. To the ATPS prepared, 0.5 mL of 10 ppm Cd(II) stock solution, 0.1 mL of 1% (w/w) of H_2SO_4, and 0–0.4 mL of 2 mol L^{-1} KI were added. The total volume of the solutions added to

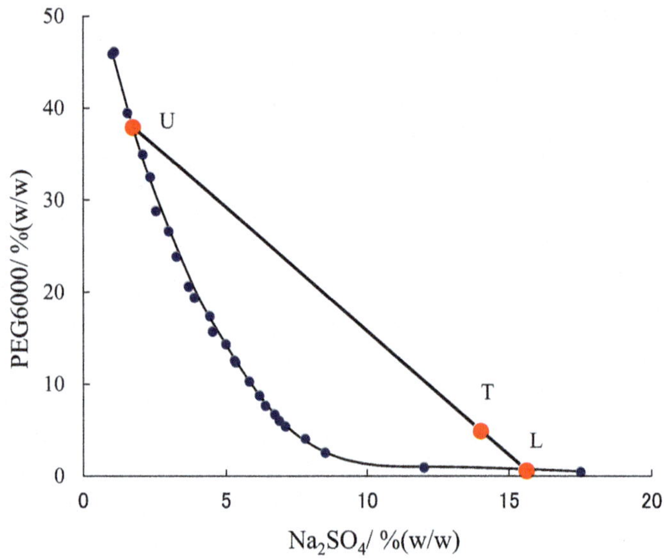

Fig. 4.8 Phase diagram for PEG6000/Na_2SO_4 ATPS. The points T, U, and L denote the total composition and the compositions of the upper and lower phases of the extraction system. Temperature: 25.0 ± 0.1 °C (reproduced from Ref. [24] by permission of Springer Nature)

the ATPS was adjusted to 2 mL with water. The points T, U, and L shown in Fig. 4.8 represent the total composition and the compositions of the upper and lower phases of the ATPS used, respectively. As can be seen from Fig. 4.9, a quantitative extraction of Cd(II) (>99%) into the upper PEG-rich phase was attained when 0.6 mmol or more of KI was added. Thus, in the following experiments, 0.8 mmol of KI was added to the extraction system.

4.3.2 Selective Extraction of Cd(II) in a Soil Sample

Extraction of metal ions from soil samples is usually performed using high concentrations of strong acids such as HCl and HNO_3. However, the ATPS extraction which uses iodide ion as an extractant allows selective extraction of Cd(II) from soil at moderate acid concentrations. This means that extraction of Cd(II) from soil and extractive separation of Cd(II) from interfering foreign metal ions in the soil samples can be achieved in a single step extraction.

The effect of the amount of H_2SO_4 on the extraction of Cd(II) in 0.1 g of the reference material of soil, JSAC 0402 (Japan Society for Analytical Chemistry, Tokyo, Japan), was examined by PEG6000/Na_2SO_4 ATPS. The ATPS was prepared by mixing 1 g of 50%(w/w) PEG6000, 7 g of 20%(w/w) Na_2SO_4, 0.4 mL of 2 mol L^{-1} KI, 0–1.5 mL of 10% (w/w) of H_2SO_4, and 0.1–1.6 mL of water (the total volume of

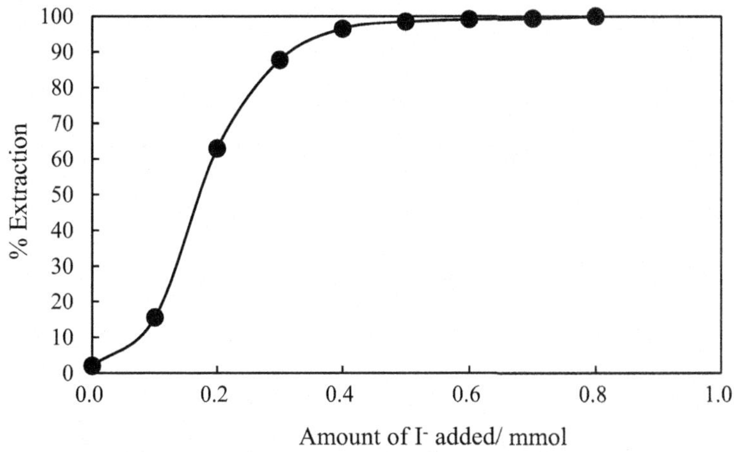

Fig. 4.9 Extraction of Cd(II) with iodide ion in PEG6000/Na_2SO_4 ATPS formed by mixing 1.0 g of 50%(w/w) PEG6000, 7.0 g of 20%(w/w) Na_2SO_4, 0.1 mL of 1%(w/w) H_2SO_4, and 1.9 mL of water containing KI. The amount of Cd(II) in the extraction system is 5 μg (reproduced from Ref. [24] by permission of Springer Nature)

10% (w/w) of H_2SO_4 and water is 1.6 mL). The mixture was shaken for 5–30 min and then centrifuged at 3000 rpm for 10 min. The upper phase was filtered with a 0.45 μm membrane filter and the concentration of Cd(II) was determined by inductively coupled plasma atomic emission spectrometry (ICP-AES). The variation of the percent extraction of Cd(II) with the amount of H_2SO_4 added to the system is represented in Fig. 4.10. Cd(II) in the soil sample is completely extracted by the addition of 1.2 mmol of H_2SO_4 to the ATPS indicating that iodide ion is a powerful extracting reagent for Cd(II). The percent extraction of Cd(II) is approximately constant over the time of extraction experiment (5–30 min). Therefore, the extraction of Cd(II) in soil and sediment samples was carried out for 30 min by adding 1.5 mmol of H_2SO_4 to the ATPS in the following experiments.

The amounts of Cd(II) and other metal ions extracted from the soil sample into the PEG-rich phase by the ATPS extraction are shown in Table 4.1. The extraction of the metal ions with 1 mol L^{-1} HCl was also examined according to the test method in Notification No.19 of the Ministry of the Environment of Japan based on the Soil Contamination Countermeasures Law. In a 50 mL polypropylene bottle, 1 g of the reference material of soil (JSAC 0402) was shaken with 35 mL of 1 mol L^{-1} HCl for 2 h. The extract was filtered with a 0.45 μm membrane filter and 1 mL of the filtrate was transferred into a 10 mL volumetric flask. After the solution was diluted to the mark with water, the concentrations of Cd, Mn, Fe, Ni, Cu, Zn, and Pb were determined by ICP-AES. The results obtained are also shown in Table 4.1. As can be seen from Table 4.1, Cd(II) is completely extracted by either the ATPS extraction with I^- or the digestion with 1 mol L^{-1} HCl, but the ATPS extraction into the PEG-rich phase of the other metal ions in the soil is significantly suppressed. This

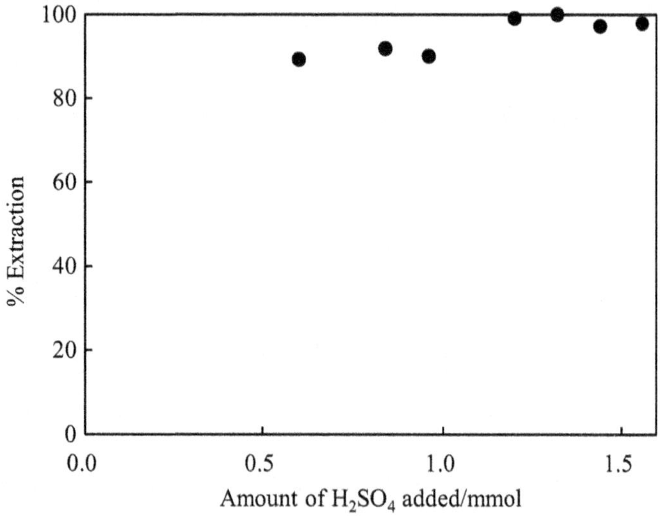

Fig. 4.10 Effect of the amount of H_2SO_4 added to the ATPS on the extraction of Cd(II) in a soil sample. The amount of soil: 0.1 g. The extraction system was prepared by mixing 1 g of 50% (w/w) PEG6000 solution, 7 g of 20% (w/w) Na_2SO_4 solution, 0.4 mL of 2 M KI solution, and 1.6 mL of water containing H_2SO_4 (reproduced from Ref. [24] by permission of Springer Nature)

indicates that the aqueous two-phase extraction with iodide ion is highly selective for the extraction of Cd(II) from soils.

The influence of the amount of soil samples on the quantification results for Cd(II) was examined. Figure 4.11 shows the dependence of the volume of the PEG-rich phase on the amount of the soil added to the ATPS. From Fig. 4.11, it can be seen that the volume of the PEG-rich phase decreases as the amount of soil increases. This may be attributed to the adsorption of the PEG molecules onto the surface of soil particles, leading to an inaccurate determination of Cd(II) if the weight or volume of the PEG-rich phase separated from the salt-rich phase and the soil is used for calculation of the amount of Cd(II) extracted. Therefore, the weight of the PEG-rich

Table 4.1 Amounts of metal ions (μg) extracted from 0.1 g of a soil sample (JSAC 0402) by the ATPS extraction with iodide ion and the digestion with hydrochloric acid (reproduced from Ref. [24] by permission of Springer Nature)

	Cd	Mn	Fe	Ni	Cu	Zn	Pb
ATPS extraction[a]	1.69	n.d.[c]	1.35	n.d.[c]	0.048	n.d.[c]	0.058
Acid digestion[b]	1.71	10.0	130	0.13	0.60	1.40	3.05

[a]Extraction with the ATPS prepared by mixing 1 g of 50% (w/w) PEG6000 solution, 7 g of 20% (w/w) Na_2SO_4 solution, 400 μL of 2 mol L^{-1} KI solution, 1.2 mL of 10% (w/w) of H_2SO_4, and 400 μL of water

[b]Digestion with 1 mol L^{-1} HCl

[c]Not determined

phase measured for the ATPS containing no soils was used for the calculation of the amount of Cd(II) extracted. The quantification results obtained for different amounts of the soil sample are given in Table 4.2. The constant value for the amount of Cd(II) extracted to the PEG-rich phase was obtained independent of the amount of soil sample within the range examined, 0.05–0.5 g of soil.

[a]Mean ± standard deviation ($n = 3$)

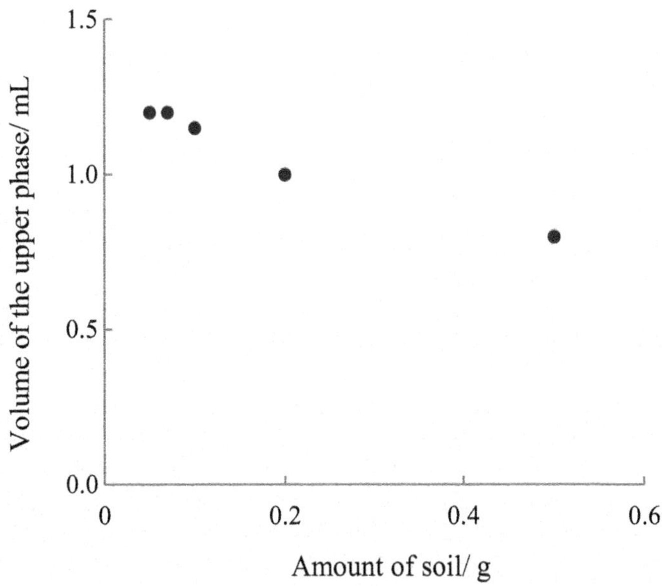

Fig. 4.11 Variation of the volume of the PEG-rich phase with the amount of the soil added to the PEG6000/Na_2SO_4 ATPS (reproduced from Ref. [24] by permission of Springer Nature)

Table 4.2 Analytical results of Cd(II) obtained for different amounts of a soil sample (JSAC 0402) (reproduced from Ref. [24] by permission of Springer Nature)

Amount of soil/g	Quantification result/μg g^{-1}[a]
0.05	17.2 ± 1.0
0.07	17.3 ± 0.5
0.1	17.1 ± 0.6
0.2	16.6 ± 0.5
0.5	16.7 ± 0.3

4.3.3 Enhancement of Selectivity of Extraction and Chromogenic Reaction of Cd(II) Using Masking Reagents

As a chromogenic reagent for the spectrophotometric determination of Cd(II), 5-(5-bromo-2-pyridylazo)-5-[*N*-*n*-propyl-*N*-(3-sulfopropyl)amino]-phenol (5-Br-PAPS) was adopted. 5-Br-PAPS is a water-soluble and a highly sensitive colorimetric reagent for transition metal ions. The absorbance of the Cd-5-Br-PAPS complex increases with an increase in pH from 7.0 to 9.0 and remains constant between 9.0 and 10.0. Therefore, the pH of the measurement solution was set at 9.7. The absorption spectrum of Cd(II)-5-Br-PAPS complex is shown in Fig. 4.12 together with those for the complexes of Ni(II), Zn(II), Cu(II), Mn(II), Hg(II), Pb(II), and Fe(III). The absorption coefficient for the Cd(II) complex at maximum absorption wavelength, 556.5 nm, is 1.15×10^5 L mol^{-1} cm^{-1}. The calibration curve is linear in the concentration range of 0–250 μg L^{-1}, and the determination limit calculated with ten times the standard deviation of the reagent blank value is 14.5 μg L^{-1}.

As shown in Table 4.1, only Cd(II) is completely extracted into the PEG-rich phase in the PEG/Na_2SO_4/KI ATPS. However, Fe(III), Cu(II), and Pb(II) are also slightly distributed into the PEG-rich phase, which may interfere with accurate determination of Cd(II), because most of the 5-Br-PAPS complexes of transition metal ions exhibit similar absorption spectra (Fig. 4.12). Since the main component of ATPS is water, conventional masking methods used in liquid–liquid extraction with organic solvents cannot directly be applied to the extraction with ATPS. It is necessary to use masking reagents that exhibit strong hydrogen bonding properties. Phosphate ion forms stable complex with Fe(III) [45] and suppresses the extraction of Fe(III) into the PEG-rich phase in PEG/salt ATPS [37]. Therefore, Na_2HPO_4 was added to strip Fe(III) from the upper phase after the ATPS extraction with H_2SO_4.

Salicylaldoxime acts as a good masking agent for Cu(II) in the complexation with 5-Br-PAPS [37]. Thus, this complexing reagent was added to the sample solution before the chromogenic reaction with 5-Br-PAPS. A solution of Na_2SO_3 was also used for suppression of oxidation of iodide ion.

It has been known that Hg(II) distributes into the PEG-rich phase in a PEG/salt ATPS with iodide ion [18, 21]. However, the complexation of Hg(II) with 5-Br-PAPS is negligible in the solution containing excess amount of iodide ion because Hg(II) forms very stable complex with iodide ion [45]. Therefore, the interference of Hg(II) with the determination of Cd(II) by the present method can be neglected.

4.3.4 Determination of Cd(II) in Soil and Sediment Samples

The standard procedure for the determination of Cd(II) in soil and sediment samples adopted is as follows: (1) about 0.1 g of a soil or sediment sample, accurately weighed, is transferred into a 15 mL polypropylene centrifuge tube; (2) to this centrifuge tube,

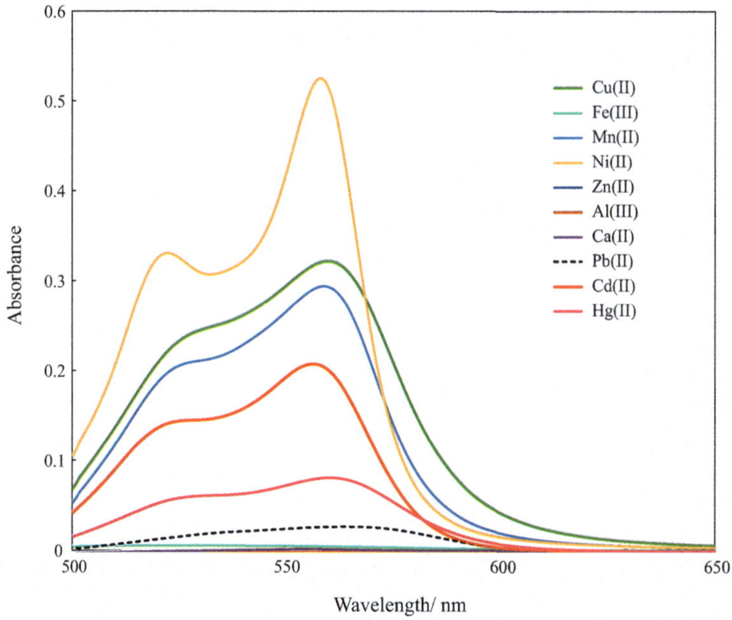

Fig. 4.12 Absorption spectra of metal complexes of 5-Br-PAPS. Concentration of 5-Br-PAPS: 30 μmol L^{-1}. Concentration of metal ions: 0.2 mg L^{-1}. pH: 9.7 (reproduced from Ref. [24] by permission of Springer Nature)

1.0 g of 50% (w/w) PEG6000, 7.0 g of 20% (w/w) Na_2SO_4, 0.4 mL of 2 mol L^{-1} KI, 1.2 mL of 10% (w/w) H_2SO_4, and 0.4 mL of water are added and the mixture is mechanically shaken for 30 min; (3) after the addition of 0.45 g of Na_2HPO_4, the mixture is again shaken for 30 min and then centrifuged for 10 min at 3000 rpm; (4) the PEG-rich phase is filtered with a 0.45 μm membrane filter and 0.5 g of the filtrate is transferred into a 10 mL volumetric flask followed by the addition of 1 mL of 0.1 mol L^{-1} Na_2SO_3, 1 mL of 1% (w/w) salicylaldoxime, 1 mL of 1% (w/w) dimethylglyoxime, and 2 mL of 1 mol L^{-1} $NaHCO_3$-Na_2CO_3 buffer (pH 9.7); (5) after the solution is allowed to stand for 20 min, 0.6 mL of 0.7 mmol L^{-1} 5-Br-PAPS is added and the solution is diluted to the mark with water; and (6) the absorbance is measured at 556.5 nm against a reagent blank. The cadmium content in a sample is determined using a calibration graph.

The effect of foreign ions on the determination of Cd(II) was examined by measuring the absorbance of the reaction mixture containing 1 μg of Cd(II) in the presence of different amounts of foreign ions. The tolerance limits for some transition metal ions are presented in Table 4.3. The criterion for interference is the value of absorbance varying by more than ±5% from the expected value.

Accuracy and precision of the method were evaluated by determining the concentrations of Cd(II) in certified reference materials for soil and river sediment. Since the PEG-rich phase was colored orange after the extraction of the river sediment

Table 4.3 Tolerance limits of foreign metal ions in determination of 1 μg of Cd(II) (reproduced from Ref. [24] by permission of Springer Nature)

Metal ion	Tolerance limit/μg
Mn(II)	30
Fe(III)	100
Ni(II)	10
Cu(II)	15
Zn(II)	100
Hg(II)	100
Pb(II)	40

sample, the absorbance was calculated by subtracting the absorbance obtained for the solution of the PEG-rich phase containing no chromogenic reagent from that of the final solution prepared according to the standard procedure. The amounts of Cd(II) in the soil and sediment samples determined by the present method are shown in Table 4.4. These values are in good agreement with the certified values, indicating that the present method gives highly reliable results. Table 4.5 shows the amounts of some transition metal ions extracted into the PEG-rich phase of the ATPS, determined by ICP-AES, in comparison with those in the sediment sample calculated from the certified values. It should be noted that the extraction of the metal ions except Cd(II) into the PEG-rich phase is almost completely suppressed.

This work describes the use of an ATPS extraction coupled to spectrophotometry to determine metal ions in soil and sediment samples and demonstrates that

Table 4.4 Determination of Cd(II) in soil and sediment reference materials (reproduced from Ref. [24] by permission of Springer Nature)

Sample[a]	Proposed method/μg g^{-1}[b]	Certified value/μg g^{-1}[b]
JSAC 0402	17.1 ± 0.6	17.3 ± 0.4
ERM-CC020	20.4 ± 0.6	20.8 ± 0.5

[a]Amount of sample: 0.1 g
[b]Mean ± standard deviation ($n = 3$)

Table 4.5 Comparison of the amounts of metal ions (μg) extracted from 0.1 g of a sediment sample (ERM-CC020) by the aqueous two-phase extraction and the metal contents (μg) in 0.1 g of the sediment (reproduced from Ref. [24] by permission of Springer Nature)

	Cd	Fe	Ni	Cu	Zn	Pb
Amount extracted[a]	2.00	0.59	0.20	6.99	0.02	0.10
Content in the sediment[b]	2.08	5100	15.8	56	203	25.5

[a]Extraction with the ATPS prepared by mixing 1 g of 50% (w/w) PEG6000 solution, 7 g of 20% (w/w) Na_2SO_4 solution, 400 μL of 2 mol L^{-1} KI solution, 1.2 mL of 10% (w/w) of H_2SO_4, and 400 μL of water, followed by stripping of Fe(III) with Na_2HPO_4. The determination of metal ions was performed by ICP-AES
[b]Calculated from the certified values

the extraction/spectrophotometry using ATPS has a unique and great potential as a simple and environmentally benign method for determination of trace metal ions in environmental samples.

4.4 Spectrophotometric Determination of Trace Amounts of Zinc in Environmental Water

4.4.1 Fast Phase Separation Using Convection

One of the drawbacks of the extractions with polymer-based ATPS is the slower phase separation due to a small difference in density between the two phases and high viscosity of the polymer-rich phase [51–53]. It is thus usually needed to use centrifugation to rapidly separate two phases of ATPS. However, the use of a centrifuge is undesirable in some cases in terms of on-site analysis. We found an interesting phenomenon, which can be used to accelerate the phase separation in ATPS; when a temperature difference is generated in the upper PEG-rich phase, convection occurs, and as shown schematically in Fig. 4.13, the micro-droplets of the lower salt-rich phase that have remained in the upper phase are quickly absorbed into the lower phase, accelerating the phase separation [37].

Phase separation operations using the convection are as follows. The sample water is added to a container of PEG and sodium sulfate and the container is immersed to hot water at 60 °C. PEG and sodium sulfate can be dissolved completely in about 1 min by gentle stirring. The container is then placed in water at room temperature for 5 min to cool. Both phases of the ATPS formed at this time are turbid. When the

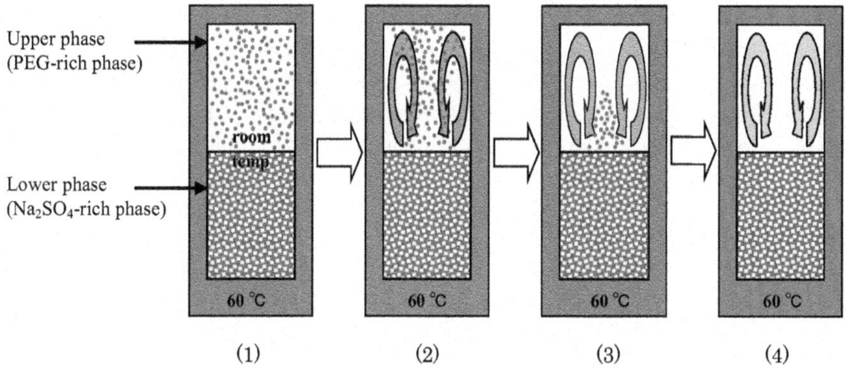

Fig. 4.13 Schematic representation of rapid phase separation of PEG/Na_2SO_4 ATPS by convection. (1): Immediately after the immersion of the sample container into the water bath; (2) and (3): The lower phase absorbs the suspended liquid particles in the upper phase accumulated by convection; (4): 2 min after the immersion of the container (reproduced from Ref. [37] by permission of The Japan Society for Analytical Chemistry)

container is then placed in hot water at 60 °C, an upward flow is generated in the upper phase near the inner wall of the container, and a convection current is generated as shown in Fig. 4.13, resulting in a downward flow in the center of the upper phase, and micro-droplets of the lower phase suspended in the upper phase are quickly absorbed into the lower phase. Although this phenomenon may be interpreted by thermal convection or Marangoni convection [54, 55], the details of the mechanism have not yet been clarified.

The greater the temperature difference between the hot and cold water, the faster the convection. When the temperature of the hot water is 60 °C and that of the cold water is room temperature, micro-droplets of the lower phase in the upper phase of the ATPS composed of 4.3%(w/w) PEG4000 and 16.8%(w/w) Na_2SO_4 can be completely removed by cooling the container of the ATPS in the cold water for 5 min and then leaving it in the hot water for 2 min. In contrast, it takes more than 2 h to completely separate the phases by allowing the system to stand at room temperature. This phenomenon is not observed in the lower phase, which remains turbid for a long time, but this is not a problem since the metal ions extracted in the upper phase are quantified.

4.4.2 *Effect of Thiocyanate Concentration on the Extraction of Zn(II)*

It was shown in Sect. 4.2 that Zn(II), Co(II), and Fe(III) are extracted by forming thiocyanate complexes into the PEG-rich phase when thiocyanate ion is added to PEG4000/Na_2SO_4 ATPS [16]. Among these metal ions, Zn(II) can be extracted into the PEG-rich phase at relatively low thiocyanate concentrations. Figure 4.14 shows the dependence of the percent extraction of Zn(II) on the concentration of thiocyanate ion in the PEG4000/Na_2SO_4 ATPS prepared by mixing 1.3 g of PEG4000, 5.1 g of anhydrous Na_2SO_4, 0.04 g of NaH_2PO_4, 0.024 g of Na_2HPO_4, and 24 mL of water containing Zn(II) at 0.1 mg L^{-1}. It can be observed that the extraction of Zn(II) is more than 95% in the thiocyanate concentration range over 0.18 mol kg^{-1}. As mentioned above, a part of the PEG-rich phase remains as micro-droplets in the lower salt-rich phase. Therefore, the actual value of the extraction is larger than 95% and the extraction of Zn(II) can be considered to be quantitative at 0.20 mol kg^{-1}. Since the volume of the PEG-rich phase is approximately 2.5 mL compared to the sample water volume of 24 mL in the PEG4000/Na_2SO_4 ATPS used, Zn(II) is concentrated approximately tenfold by this extraction.

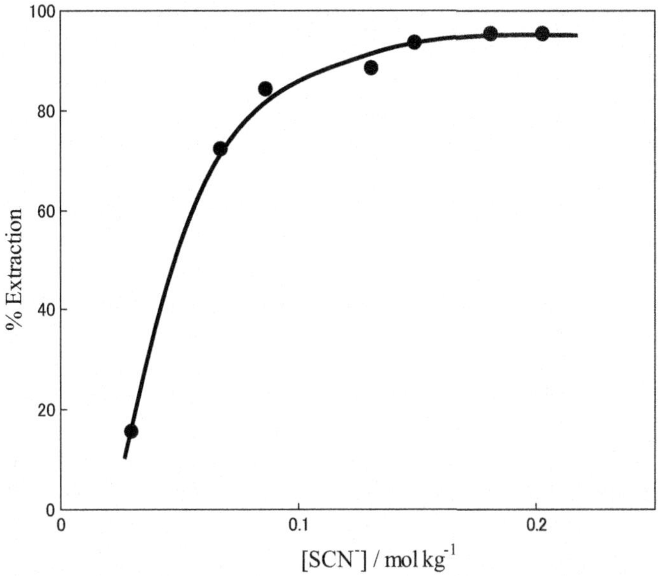

Fig. 4.14 Dependence of the extraction of Zn(II) on the concentration of thiocyanate ion in PEG4000/Na_2SO_4 ATPS. The ATPS was prepared by dissolving 1.3 g of PEG4000, 5.1 g of anhydrous Na_2SO_4, 0.04 g of NaH_2PO_4, 0.024 g of Na_2HPO_4, and 24 mL of water containing Zn(II) at 0.1 mg L^{-1} (reproduced from Ref. [37] by permission of The Japan Society for Analytical Chemistry)

4.4.3 Masking of Foreign Metal Ions for Selective Spectrophotometric Determination of Zn(II)

5-Br-PAPS was used as a chromogenic reagent for the determination of Zn(II) in environmental waters. As described in the preceding sections, thiocyanate ion extracts not only Zn(II) but also Cu(II) and Fe(III) into the PEG-rich phase in the PEG/Na_2SO_4 ATPS (Fig. 4.2) and 5-Br-PAPS forms red chelates with all of these metal ions (Fig. 4.12). Therefore, a method to suppress the extraction of Cu(II) and Fe(III) into the PEG-rich phase was investigated. As a result, it was found that the addition of phosphate ion and thiosulfate ion to the extraction system suppressed the extraction of Fe(III) and Cu(II) into the upper PEG-rich phase, respectively [37]. Both of these anions form kosmotropic or water structure-making anionic metal complexes and keep Fe(III) and Cu(II) in the lower salt-rich phase.

In the ATPS used, most metal ions do not partition into the PEG-rich phase, and the distribution ratios of metal ions complexing with 5-Br-PAPS other than Zn(II), Cu(II), and Fe(III) are less than 0.03. Therefore, if the concentrations of foreign metal ions coexisting in the sample water are less than 10 times the concentration of Zn(II), then they will not interfere with the determination of Zn(II). However, in addition to Fe(III) and Cu(II), Co(II), Ni(II), and Mn(II) may be present in environmental water at concentrations similar to that of Zn(II) as shown below. Therefore, except for Fe(III), which can be completely removed in the extraction process, the masking of these metal ions was investigated to remove their interference with the color reaction of Zn(II).

A number of reagents have been reported as effective agents to mask these metal ions. Among them, salicylaldoxime [56] and dimethylglyoxime [57] were found to be appropriate masking agents for Cu(II) and Ni(II), respectively. However, a gradual increase in absorbance was observed when 5-Br-PAPS was added after dimethylglyoxime and a pH buffer solution was added to the solution of Ni(II). This is presumably due to the slow conversion of Ni(II)-dimethylglyoxime complex to the 5-Br-PAPS complex. In an experiment with 1 mg L^{-1} Ni(II) coexisting with 0.1 mg L^{-1} Zn(II), the absorbance does not change within 10 min after the addition of 5-Br-PAPS at room temperature. The absorbance of the solution was thus measured immediately after 5-Br-PAPS was added and the solution was stirred.

It was found that Mn(II) can be masked by adding ammonium peroxodisulfate and triethanolamine to form a triethanolamine complex of Mn(III). On the other hand, Co(II) was masked by the following reactions to form a Co(III)-ethylenediamine complex:

$$Co^{2+} + 3en \rightarrow Co(en)_3^{2+}$$

$$2Co(en)_3^{2+} + S_2O_8^{2-} \rightarrow 2Co(en)_3^{3+} + 2SO_4^{2-}$$

where en denotes ethylenediamine. Although this reaction takes 20 min to complete at room temperature, 5 min is sufficient at 60 °C. Therefore, ethylenediamine and ammonium peroxodisulfate were added to the upper phase after the ATPS extraction, and then 5-Br-PAPS was added after the sample container was soaked in hot water at 60 °C for 5 min. Zn(II) also forms a complex with ethylenediamine, but it is rapidly converted to the complex with 5-Br-PAPS because the formation constant of Zn(II)-5-Br-PAPS complex is larger than that of the complex with ethylenediamine and the substitution reaction is very fast.

4.4.4 Determination of Zn(II) in Tap Water and River Water Samples

The procedure for the determination of Zn(II) in environmental water samples is as follows: (1) in a 25 mL volumetric flask, 1.3 g of PEG4000, 5.1 g of anhydrous sodium sulfate, 0.46 g of potassium thiocyanate, 0.18 g of sodium thiosulfate pentahydrate, 0.04 g of sodium dihydrogen phosphate, and 0.024 g of disodium hydrogen phosphate are placed, to which 24 mL of sample water is added; (2) the volumetric flask is placed in a water bath at 60 °C and the contents in the flask are shaken to dissolve; (3) the volumetric flask is placed in cold water (at room temperature) for 5 min, and then immersed again in a water bath at 60 °C for 2 min; (4) a 2.00 g aliquot of the separated upper phase is pipetted into a 10 mL test tube containing 0.023 g of ammonium peroxodisulfate, to which is added 3.9 mL of a 66% aqueous ethanol solution containing the masking agents; the ethanol solution is prepared by mixing 0.024 g dimethylglyoxime, 0.012 g salicylaldoxime, 0.67 μL ethylenediamine, 0.1 g triethanolamine, 2.5 mL ethanol, 1.3 mL water, and 252 μL of 4 mol L^{-1} sodium hydroxide solution; (5) the test tube is immersed in a water bath at 60 °C, allowed to stand for 5 min, and then placed in cold water (room temperature) to cool; (6) to this solution, 40 μL of 7.65 mol L^{-1} 5-Br-PAPS solution was added, and the absorbance was measured at 552 nm.

The calibration curve was prepared using aqueous solutions containing 0–200 μg L^{-1} Zn(II) according to the method described above. The calibration curve obtained shows good linearity ($r^2 = 0.9996$) as shown in Fig. 4.15, and the determination limit calculated with ten times the standard deviation of the reagent blank value is 1.6 μg L^{-1}.

The influences of foreign ions on the determination of Zn(II) were examined by measuring the absorbance of the reaction mixture containing 10 μg of Zn(II) in the presence of different amounts of foreign ions. The tolerance limits taken as the amount that causes ±5% error in the absorbance value are presented in Table 4.6. The presence of Fe(II) in the concentration level up to 200-fold to the concentration of Zn(II) has negligible influence on the absorbance. This is presumably due to the oxidation of Fe(II) to Fe(III) by ammonium peroxodisulfate and subsequent complexation with triethanolamine, which were added for masking of Co(II). The

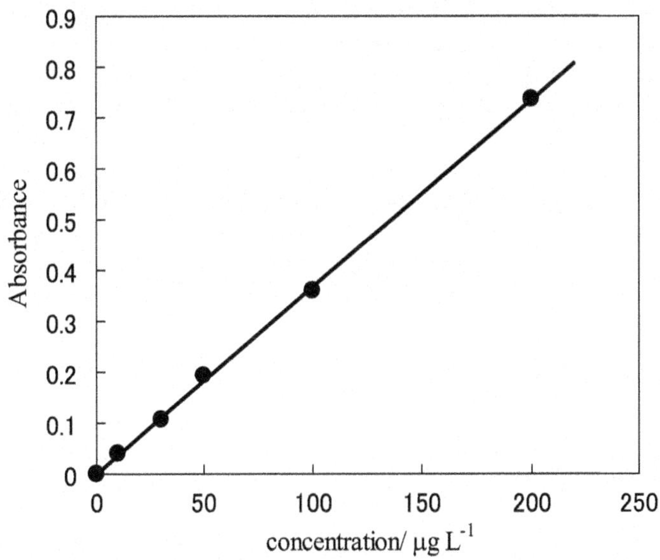

Fig. 4.15 Calibration curve for the determination of Zn(II) by the ATPS extraction-spectrophotometric method (reproduced from Ref. [37] by permission of The Japan Society for Analytical Chemistry)

Table 4.6 Effect of foreign ions on the determination of 10 μg L^{-1} Zn(II) (reproduced from Ref. [37] by permission of The Japan Society for Analytical Chemistry)

Added ion	Tolerance limit/mg L^{-1}
Cu^{2+}	0.3
Ni^{2+}	0.8
Mn^{2+}, Fe^{2+}	2.0
Co^{2+}	5.0
Fe^{3+}	20
Na^{+}, Mg^{2+}, Ca^{2+}	>100
Cl^{-}	500
CH_3COO^{-}	>1000

tolerance limit of Fe(III) by masking as a phosphate complex in the extraction stage alone was 2 mg L^{-1}, whereas the determination of Zn(II) by the procedure finally adopted was not affected by the coexistence of 20 mg L^{-1} of Fe(III). This may also be due to the masking effect of triethanolamine.

In addition to metal ions, the interference of chloride and acetate ions, which are present in relatively large amounts in environmental water and have the potential to form complexes with Zn(II), was examined. The coexistence of acetate ion at a concentration of 1000 mg L^{-1} has no effect, while that of chloride ion in the sample water at the same concentration results in a negative error of 87% recovery in the determination of Zn(II) at 10 μg L^{-1}, which may be attributed to the complexation of Zn(II) with chloride ion.

The present ATPS extraction-spectrophotometric method was applied to the analysis of Zn(II) in river water samples collected from Ara River and Yanase River located in Saitama City, Japan, and tap water sampled at our laboratory. All of the water samples were filtered through a membrane filter (0.45 μm) after sampling. The filtered river water samples were adjusted to pH 3 with 1 mol L^{-1} sulfuric acid. The analytical results obtained are shown in Table 4.7. The results obtained for Zn, Mn, Fe, Co, Ni, and Cu by ICP-AES are also shown in Table 4.8 for comparison. The concentration of Zn(II) in the tap water sample could not be determined because it was below the determination limit of the present method. However, the results for the river water samples are in good agreement with those obtained by ICP-AES. These results indicate that the proposed ATPS extraction-spectrophotometric method allows simple on-site analysis of Zn(II) in environmental water samples.

4.5 Conclusions

The distribution behavior of Mn(II), Fe(III), Co(II), Ni(II), Cu(II), Zn(II), Cd(II), and Li(I) in PEG/Na_2SO_4 ATPS was studied using thiocyanate or iodide ion as extracting agents. From the distribution ratios of the metal ions determined as a function of iodide or thiocyanate concentration, the extracted metal species were assumed. In

Table 4.7 Analytical results obtained by the ATPS extraction-spectrophotometric method for determination of Zn(II) in tap water and river water samples (reproduced from Ref. [37] by permission of The Japan Society for Analytical Chemistry)

Sample	Zn(II) added/μg L^{-1}	Zn(II) found/μg L^{-1a}
Tap water	0	n. d.[b]
Ara River	0	22 ± 2
	10	33 ± 2
	20	45 ± 2
Yanase River	0	25 ± 1
	10	35 ± 3
	20	43 ± 2

[a]Mean ± standard deviation ($n = 3$). [b]not determined

Table 4.8 Analytical results obtained by ICP-AES for determination of some metal ions in tap water and river water samples (reproduced from Ref. [37] by permission of The Japan Society for Analytical Chemistry)

Sample	Metal ion	Concentration/μg L^{-1a}
Tap water	Zn	1.5 ± 0.1
	Mn	n. d.[b]
	Fe	1.1 ± 0.2
	Co	0.3 ± 0.3
	Ni	1.0 ± 0.2
	Cu	0.3 ± 0.3
Ara River	Zn	22.9 ± 0.1
	Mn	26.7 ± 0.1
	Fe	35.1 ± 0.2
	Co	n. d.[b]
	Ni	5.0 ± 0.4
	Cu	0.5 ± 0.6
Yanase River	Zn	26.5 ± 0.5
	Mn	16.5 ± 0.1
	Fe	49.2 ± 0.6
	Co	n. d.[b]
	Ni	4.9 ± 0.3
	Cu	7.5 ± 0.4

[a]Mean ± standard deviation ($n = 3$). [b]not determined

the PEG/Na_2SO_4/SCN^- ATPS, Cu(II), Zn(II), Co(II), and Fe(III) distribute into the PEG-rich phase as four-coordinated thiocyanate complexes. On the other hand, Cd(II) is extracted as $Cd(SCN)_2$ in the ATPS with SCN^-, while as CdI_4^{2-} in the I^- system. The extractabilities of the metal ions depend not only on the formation constants of the metal thiocyanate or iodide complexes but also on those of their sulfate complexes. Sulfate ion prevents the distribution of metal ions into the PEG-rich phase by formation of sulfate complexes in the Na_2SO_4-rich phase. This indicates that the extractability of a metal ion with complexing anions in ATPS formed by a polymer and an inorganic salt depends not only on the complexing ability of the extracting agents but also that of the anion of the phase-forming salt.

The PEG/salt ATPSs are useful for selective extraction of Cd(II) with iodide ion and of Zn(II) with thiocyanate ion as the extracting agents and can be used for simple spectrophotometric determination of trace amounts of these metal ions in environmental samples. The chromogenic reaction of 5-Br-PAPS is not selective for a particular metal ion but it forms complexes with many metal ions and exhibits very high absorptivities. However, the coupling of the selective ATPS extraction procedure with spectrophotometry makes it possible to determine accurately the analyte metal ions in various samples.

A rapid phase separation technique for the ATPS was developed using convection so that the extraction can be accomplished within 7 min without mechanical centrifugation. The ATPS extraction method developed is more environmentally friendly than the usual ones resorting to organic solvents and, furthermore, enables one to selectively extract the target metal ion in not only river water samples but also soil and sediment samples. The method provides a universal protocol for the extraction-spectrophotometric determination of various metal ions in environmental samples and is expected to be readily applied to other metal ions of interest using the reagents which selectively extract them.

References

1. Rogers RD, Eiteman MA (eds) (1995) Aqueous biphasic separations: biomolecules to metal ions. Plenum Press, New York
2. Karmakar R, Sen K (2019) Aqueous biphasic extraction of metal ions: an alternative technology for metal regeneration. J Mol Liq 273:231–247
3. Zvarova TI, Shkinev VM, Vorob'eva GA, Spivakov BY, Zolotov YA (1984) Liquid-liquid extraction in the absence of usual organic solvents: application of two-phase aqueous systems based on a water-soluble polymer. Microchim Acta 84:449–458
4. Molochnikova NP, Shkinev VM, Myasoedov BF (1992) Two-phase aqueous systems based on poly(ethylene glycol) for extraction of actinides in various media. Solvent Extr Ion Exc 10:697–712
5. Rogers RD, Bond AH, Bauer CB (1993) The crown ether extraction of group 1 and 2 cations in polyethylene glycol-based aqueous biphasic systems at high alkalinity. Pure Appl Chem 65:567–572
6. Rogers RD, Bond AH, Bauer CB (1993) Aqueous biphase systems for liquid/liquid extraction of f-elements utilizing polyethylene glycols. Sep Sci Technol 28:139–153
7. Rogers RD, Bond AH, Bauer CB (1993) Metal ion separations in polyethylene glycol-based aqueous biphasic systems. Sep Sci Technol 28:1091–1126
8. Rogers RD, Bond AH, Bauer CB, Zhang J, Griffin ST (1996) Metal ion separations in polyethylene glycol-based aqueous biphasic systems: correlation of partitioning behavior with available thermodynamic hydration data. J Chromatogr B 680:221–229
9. Rogers RD, Zhang J (1996) Effects of increasing polymer hydrophobicity on distribution ratios of TcO_4^- in polyethylene/poly(propylene glycol)-based aqueous biphasic systems. J Chromatogr B 680:231–236
10. Rogers RD, Bauer CB (1996) Partitioning behavior of group 1 and 2 cations in poly(ethylene glycol)-based aqueous biphasic systems. J Chromatogr B 680:237–241

11. Rogers RD, Griffin ST (1998) Partitioning of mercury in aqueous biphasic systems and on ABEC™ resins. J Chromatogr B 711:277–283
12. Spear SK, Griffin ST, Huddleston JG, Rogers RD (2000) Radiopharmaceutical and hydrometallurgical separations of perrhenate using aqueous biphasic systems and the analogous aqueous biphasic extraction chromatographic resins. Ind Eng Chem Res 39:3173–3180
13. Chen X, Li G, Hu Z (1996) Determination of microamounts of palladium(II) by extraction method without usual organic solvents. Mikrochim Acta 122:143–149
14. Lamb JD, Nazarenko AY, Neilson L (1998) Novel extraction of metal ions from acidic media using poly(ethylene oxide) phases. Anal Commun 35:145–146
15. Pai SA, Lohithakshan KV, Mithapara PD, Aggarwal SK (2000) Use of polyethylene glycol (PEG) based aqueous biphasic system (ABS) for the extraction of Pu(IV), Pu(VI) and Am(III) with 18-crown-6: a thermodynamic study. J Radioanal Nucl Chem 245:619–622
16. Shibukawa M, Nakayama N, Hayashi T, Shibuya D, Endo Y, Kawamura S (2001) Extraction behaviour of metal ions in aqueous polyethylene glycol-sodium sulphate two-phase systems in the presence of iodide and thiocyanate ions. Anal Chim Acta 427:293–300
17. Yoshikuni N, Baba T, Tsunoda N, Oguma K (2005) Aqueous two-phase extraction of nickel dimethylglyoximato complex and its application to spectrophotometric determination of nickel in stainless steel. Talanta 66:40–44
18. Bulgariu L, Bulgariu D (2006) Hg(II) extraction in a PEG-based aqueous two-phase system in the presence of halide ions. I. Liquid phase analysis. Central Eur J Chem 4:246–257
19. Bulgariu L, Bulgariu D, Sârghie I, Măluţan T (2007) Cd(II) extraction in PEG-based two-phase aqueous systems in the presence of iodide ions. Analysis of PEG-rich solid phases. Central Eur J Chem 5:291–302
20. Bulgariu L, Bulgariu D (2008) Extraction of metal ions in aqueous polyethylene glycol–inorganic salt two-phase systems in the presence of inorganic extractants: correlation between extraction behaviour and stability constants of extracted species. J Chromatogr A 1196–1197:117–124
21. Bulgariu L, Bulgariu D (2013) Selective extraction of Hg(II), Cd(II) and Zn(II) ions from aqueous media by a green chemistry procedure using aqueous two-phase systems. Sep Purif Technol 118:209–216
22. Roy K, Lahiri S (2008) Species dependent radiotracer study of Cr(VI) and Cr(III) using an aqueous biphasic system. Radiochim Acta 96:49–53
23. Sui N, Huang K, Zhang C, Wang N, Wang F, Liu H (2013) Light, middle, and heavy rare-earth group separation: a new approach via a liquid-liquid-liquid three-phase system. Ind Eng Chem Res 52:5297–6008
24. Shibukawa M, Nobushima D, Sakuma S, Sasaki M, Nakamura K, Matsutani Y, Saitoh K, Saito S (2016) Selective spectrophotometric determination of trace amounts of cadmium in soil and sediment samples using a green aqueous biphasic extraction. Anal Sci 32:1095–1100
25. de Oliveira WCM, Rodrigues GD, Mageste AB, de Lemos LR (2017) Green selective recovery of lanthanum from Ni-MH battery leachate using aqueous two-phase systems. Chem Eng J 322:346–352
26. Hamta A, Dehghani MR (2017) Application of polyethylene glycol based aqueous two-phase systems for extraction of heavy metals. J Mol Liq 231:20–24
27. Sun P, Huang K, Lin J, Liu H (2018) Role of hydrophobic interaction in driving the partitioning of metal ions in a PEG-based aqueous two-phase system. Ind Eng Chem Res 57:11390–11398
28. Patricio PR, Mesquita MC, da Silva LHM, da Silva MCH (2011) Application of aqueous two-phase systems for the development of a new method of cobalt(II), iron(III) and nickel(II) extraction: a green chemistry approach. J Hazard Mater 193:311–318
29. de Lemos LR, Santos IJB, Rodrigues GD, da Silva LHM, da Silva MCH (2012) Copper recovery from ore by liquid-liquid extraction using aqueous two-phase system. J Hazard Mater 237–238:209–214
30. Patrício PR, Cunha RC, Vargas SJR, Coelho YL, da Silva LHM, da Silva MCH (2016) Chromium speciation using aqueous biphasic systems: Development and mechanistic aspects. Sep Purif Technol 158:144–154

31. Patrício PR, Quintão JC, da Silva LHM, Hespanhol MC (2022) One-pot separation and preconcentration for the determination of chromium species using an aqueous biphasic system. J Braz Chem Soc 33:978–986
32. Shirayama S, Uda T (2015) Recovery of cobalt ion into polyethyleneglycol (PEG) gel phase as thiocyanato complex. Mater Trans 56:610–616
33. Leite DS, Carvalho PLG, de Lemos LR, Mageste AB, Rodrigues GD (2017) Hydrometallurgical separation of copper and cobalt from lithium-ion batteries using aqueous two-phase systems. Hydrometallurgy 169:245–252
34. Umebayashi Y, Ishiguro S (1996) Spectrophotometric study of thiocyanato complexation of cobalt(II) and nickel(II) ions in micellar solutions of a nonionic surfactant Triton X-100. J Solut Chem 25:731–746
35. Umebayashi Y, Shin M, Kanzaki R, Ishiguro S (2001) Thermodynamics of $[Co(NCS)_4]^{2-}$ at poly(ethylene oxide) and octylphenyl moieties in micelles of nonionic surfactants. J Colloid Interf Sci 237:167–173
36. Marcus Y (2009) Effect of ions on the structure of water: structure making and breaking. Chem Rev 109:1346–1370
37. Shibukawa M, Shirota D, Saito S, Nagasawa S, Saitoh K, Minamisawa H (2010) Simple spectrophotometric determination of trace amounts of zinc in environmental water samples using aqueous biphasic extraction. Bunseki Kagaku 59:847–854
38. Malik LA, Bashir A, Qureashi A, Pandith AH (2019) Detection and removal of heavy metal ions: review. Environ Chem Lett 17:1495–1521
39. Nagajyoti PC, Lee KD, Sreekanth EVM (2010) Heavy metals, occurrence and toxicity for plants: a review. Environ Chem Lett 8:199–216
40. Haddad JE, Villot-Kadri M, Ismaël A, Gallou G, Michel K, Bruyère D, Laperche V, Canioni L, Bousquet B (2013) Artificial neural network for on-site quantitative analysis of soils using laser induced breakdown spectroscopy. Spectrochim Acta Part B 79–80:51–57
41. Ahmed MJ, Chowdhury MTI (2004) A simple spectrophotometric method for the determination of cadmium in industrial, environmental, biological and soil samples using 5,7-dibromo-8-hydroxyquinoline. Anal Sci 20:987–990
42. Reddy DN, Reddy KV, Reddy TS, Reddy KH (2011) Extractive spectrophotometric determination of trace amounts of cadmium (II) in medicinal leaves environmental and biological samples by using 3-methylthiophene-2-carbaxaldehyde thiosemicarbazone (3-MTAT). Adv Appl Sci Res 2(4):328–337
43. El-Zomrawy AA (2018) Selective and sensitive spectrophotometric method to determine trace amounts of copper metal ions using Amaranth food dye. Spectrochim Acta Part A 203:450–454
44. Shibukawa M, Matsuura K, Shinozuka Y, Mizuno S, Oguma K (2000) Effects of phase-forming cations and anions on the partition of ionic solutes in aqueous polyethylene glycol-inorganic salt two-phase systems. Anal Sci 16:1039–1044
45. Sillén LG, Martell AE (1964) Stability constants of metal-ion complexes. The Chemical Society, London
46. Piscator M (1985) Dietary exposure to cadmium and health effects: impact of environmental changes. Environ Health Perspect 63:127–132
47. Chibuike GU (2014) Obiora SC (2014) Heavy metal polluted soils: effect on plants and bioremediation methods. Appl Environ Soil Sci 1:752708
48. Kubier A, Wilkin RT, Pichler T (2019) Cadmium in soils and groundwater: a review. Appl Geochem 108:104388
49. Reddy SA, Reddy KJ, Lakshminaraya S, Priya DL, Rao YS, Reddy AV (2008) Extractive spectrophotometric determination of trace amounts of cadmium(II) in medicinal leaves and environmental samples using benzildithiosemicarbazone (BDTSC). J Hazard Mater 152:903–909
50. Bagheri A, Behbahani M, Amini MM, Sadeghi O, Taghizade M, Baghayi L, Salarian M (2012) Simultaneous separation and determination of trace amounts of Cd(II) and Cu(II) in environmental samples using novel diphenylcarbazide modified nanoporous silica. Talanta 89:455–461

51. Shibukawa M, Ichikawa R, Baba T, Sakamoto R, Saito S, Oguma K (2008) Separation selectivity of aqueous polyethylene glycol-based separation systems: DSC, LC and aqueous two-phase extraction studies. Polymer 49:4168–4173
52. Albertsson PA (1986) Partitioning of cell particles and macromolecules, 3rd edn. Wiley-Interscience, New York
53. Srinivas ND, Barhate RS, Raghavarao KSMS, Todd P (2000) Acoustic field assisted enhanced demixing of aqueous two-phase systems. Biochim Biophys Acta 1524:38–44
54. Xiao Y, Ribe NM, Zhang Y, Pan Y, Cao Y, Shun HC (2022) Generation of Fermat's spiral patterns by solutal Marangoni-driven coiling in an aqueous two-phase system. Nat Commun 13:7206
55. Palizdan S, Abbasi J, Riazi M, Malayeri MR (2020) Impact of solutal Marangoni convection on oil recovery during chemical flooding. Pet Sci 17:1298–1317
56. Lucia M, Campos AM, van den Berg CMG (1994) Determination of copper complexation in sea water by cathodic stripping voltammetry and ligand competition with salicylaldoxime. Anal Chim Acta 284:481–496
57. Qu F, Zhang Q, You J (2015) Recognition and determination of multi-metal ions based on silver nanoclusters capped by polyethyleneimine with different molecular weights. New J Chem 39:9293–9298

Chapter 5
Multistep pH-Peak-Focusing Countercurrent Chromatography

Abstract Countercurrent chromatography (CCC) is a support-free liquid–liquid chromatography using two immiscible liquid phases. ATPS has been used as a CCC platform mainly for the separation and purification of biological materials. CCC with ATPS also has great potential as a chromatographic method that allows simultaneous separation and enrichment of metal ions through pH-peak-focusing technique. This chapter begins by describing the optimum operating conditions for a type-J coil planet CCC with ATPSs consisting of PEG and Na_2SO_4. It then proceeds to explain the multistep pH-peak-focusing CCC developed for the separation and enrichment of metal ions with a PEG/Na_2SO_4 ATPS. The resolution of a type-J coil planet CCC increases with an increase in the rotation speed of the polytetrafluoroethylene tube column holder due to a drastic increase in the theoretical plate number. This is caused by the increase in the number of the partition processes and the decrease in the size of the phase droplets due to more intense agitation, which increases the mass transfer rate of solute compounds between the mobile phase (salt-rich phase) and the stationary phase (PEG-rich phase). On the other hand, the increase in the flow rate of the mobile phase does not appreciably affect the resolution of CCC. In multistep pH-peak-focusing CCC, metal ions in a sample solution are trapped at the top of the CCC column filled with a basic stationary phase containing an extraction reagent by complexation with it. Then, the mobile phases of which the pH values have been adjusted with suitable buffer reagents are pumped into the column stepwise in order of decreasing pH. Each metal ion is concentrated at a pH border formed between the zones of different pH in the CCC column through extraction as the metal complexes with an extracting agent into the stationary phase at the front side (higher pH) of the border and back extraction by the dissociation of the complexes into the mobile phase at the rear side (lower pH) of the border, eluting from the column together with the pH border. Enrichment as well as mutual separation of metal ions can be achieved by multistep pH-peak-focusing CCC. The mechanism of formation of the pH borders and the potential of multistep pH-peak-focusing CCC for preparative-scale metal separation are also discussed.

M. Shibukawa, *Polymer-Based Aqueous-Two Phase Systems*,
https://doi.org/10.1007/978-981-96-2888-9_5

Keywords Aqueous two-phase system · Countercurrent chromatography · Multistep pH-peak-focusing · Separation and enrichment of metal ions · Rare earth elements

5.1 Introduction

It is sometimes difficult to achieve selective separation of a target compound from the coexisting ones in a sample solution by liquid–liquid extraction with a conventional batch procedure. A complete separation of the target compound and interferents is possible with a sufficient number of extractions or chromatographic techniques. Countercurrent chromatography (CCC) is a liquid chromatographic technique having support-free liquid stationary phase [1–4]. It is thus relatively easy to introduce ATPS into CCC as a separation platform and a number of studies associating CCC with ATPS have been reported in the literature [2, 5–10].

CCC with ATPS has mainly been used for the separation and purification of biological substances [5–10] and there are relatively few reports of this technology being applied to the separation of inorganic compounds [11–13]. However, CCC using PEG/salt ATPS has great potential not only as a green sustainable separation method but also as a technique that allows simultaneous separation and concentration of metal ions through multistep pH-peak-focusing [12].

Kitazume et al. developed a method for highly efficient enrichment of metal ions by pH-peak-focusing CCC [14, 15]. pH-peak-focusing is a concentration technique based on neutralization between the mobile and stationary phases with pH values different from each other [12, 14–18]. Kitazume et al. used an organic solvent (ether or *n*-heptane) containing 0.22 mol L^{-1} bis-(2-ethylhexyl) phosphoric acid (DEHPA) and 0.2 mol L^{-1} ammonia as the stationary phase and 0.1 mol L^{-1} HCl as the mobile phase. When metal ions are introduced into the column, they form DEHPA complexes by contact with the basic stationary phase containing the complexing reagent and distribute into the stationary phase. Following sample injection, the acidic mobile phase is delivered to the column, and therefore neutralization reaction between HCl in the mobile phase and NH_3 in the stationary phase continuously proceeds. This neutralization reaction forms a pH border, where the pH value steeply changes. The pH border moves through the column at a lower velocity than the velocity of the mobile phase, determined by the relative amounts of HCl in the mobile phase and NH_3 in the stationary phase. As the sharp pH border moves forward, the metal complexes are exposed to low pH, dissociated, and released metal ions distribute back to the mobile phase, where they migrate at the velocity of the mobile phase. Consequently, metal ions are trapped within the pH border as depicted in Fig. 5.1 and are eluted as a sharp peak. Kitazume et al. demonstrated that metal ions injected as a 10 mL solution are concentrated over 100-fold by this technique [14].

Liquid–liquid extraction of metal ions with organic/aqueous two-phase systems is conventionally performed using an organic weak acid that forms metal complexes as a chelating ligand such as DEHPA. The extraction process includes the following

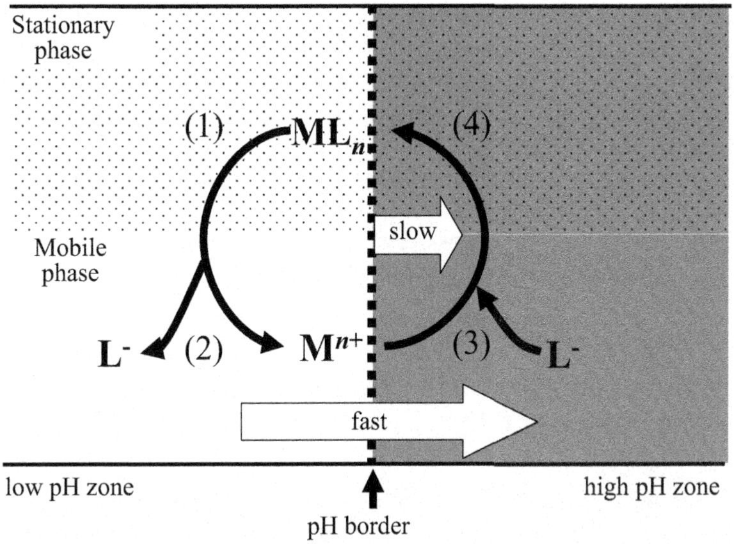

Fig. 5.1 Mechanism of pH-peak-focusing CCC. (1) the stationary phase equilibrated with the acidic mobile phase; (2) the acidic mobile phase before neutralization; (3) the neutralized mobile phase; (4) the basic stationary phase; M^{n+}, metal ion; L^-, complexing ligand; ML_n, metal complex (reproduced from Ref. [12] by permission of The American Chemical Society)

equilibria:

$$HL \rightleftarrows H^+ + L^- \tag{5.1}$$

$$M^{n+} + nL^- \rightleftarrows ML_n \tag{5.2}$$

$$ML_n \rightleftarrows \overline{ML_n} \tag{5.3}$$

where HL, M^{n+}, and ML_n are the chelating ligand, metal ion, and metal complex, respectively, and overbar denotes the organic phase. Therefore, each metal ion is extracted into the organic phase at different pH from each other according to the complexation formation constant. The larger the formation constant, the pH at which the metal ion can be extracted is smaller. It is thus expected that the pH-peak-focusing technique can be used not only for enrichment but also for separation of metal ions by forming multistep pH borders of different pHs corresponding to the extraction of target metal ions.

In order to form a pH border where the pH jump is precisely controlled in both the mobile and stationary phases, it is necessary to utilize a buffer. Since either acid form or base form of a buffering agent is ionic (protonated or deprotonated) and is insoluble in hydrophobic organic solvents, organic/aqueous two-phase systems are not suitable for constructing multistep pH borders in CCC columns. On the contrary, both of the acid and base forms of a buffering agent can distribute between the two phases of ATPS, because water is the main component of the ATPS, which means that ATPS is an ideal extraction system that fulfills the important requirement for performing multistep pH-peak-focusing CCC.

In this chapter, multistep pH-peak-focusing CCC with PEG/Na_2SO_4 ATPS, a novel eco-friendly method that can simultaneously separate and enrich metal ions, is described. It will be shown that ideal pH borders can be formed by controlling the distribution of acid and base forms of buffering agents and mutual separation of rare earth elements can be achieved by this method.

5.2 Effect of the Operating Conditions on the Separation Performance of CCC

CCC experiments were carried out using a type-J countercurrent chromatograph shown in Fig. 5.2. The column holder revolves horizontally around the vertical axis of the centrifuge, and an adjustable counter weight was mounted on the opposite side of the column holder to balance the centrifuge system. The CCC column is a polytetrafluoroethylene (PTFE) tube with an inner diameter of 2.0 mm wrapped around the hub of a holder, and the inner volume of the tube is 60.5 mL.

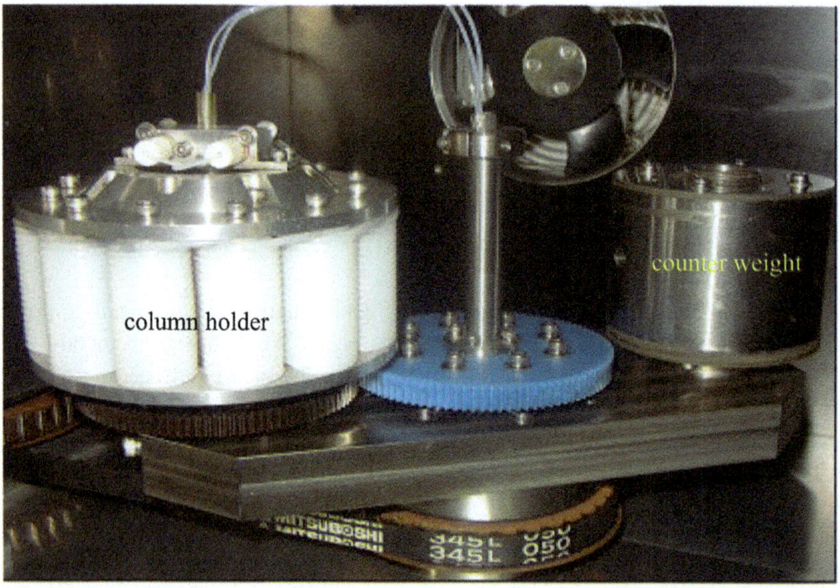

Fig. 5.2 A type-J countercurrent chromatograph

Figure 5.3 shows the phase diagrams of the PEG1000/Na_2SO_4 and PEG4000/Na_2SO_4 ATPSs used for the investigation of the effect of the CCC operating conditions on the separation efficiency. The two-phase systems were formed with the total compositions indicated by points A and B in Fig. 5.3, and the PEG-rich phase and the Na_2SO_4-rich phase were used as the stationary phase and the mobile phase, respectively. The column was first filled with the stationary phase and then the mobile phase was delivered into the column.

Examples of separations of IO_3^-/I^- and NO_3^-/SCN^- obtained using the ATPS indicated by point B in Fig. 5.3 are shown in Fig. 5.4. For hydrodynamic CCC, the constant gravitational field is augmented with a revolutionary rotation around the axis of the coil [1, 4]. Figure 5.5 shows the effect of column rotation speed on the resolution, R_s, for the separation of IO_3^- and I^- when the ATPSs of which the total compositions are shown by the points A and B in Fig. 5.3 are used. The resolution increases with an increase in the column rotation speed. The resolution for two adjacent peaks of compounds 1 and 2 can be given by

$$R_s = \frac{1}{4}\frac{(\alpha - 1)}{\alpha}N^{1/2}\left(\frac{D_2}{D_2 + V_m/V_s}\right) \quad (5.4)$$

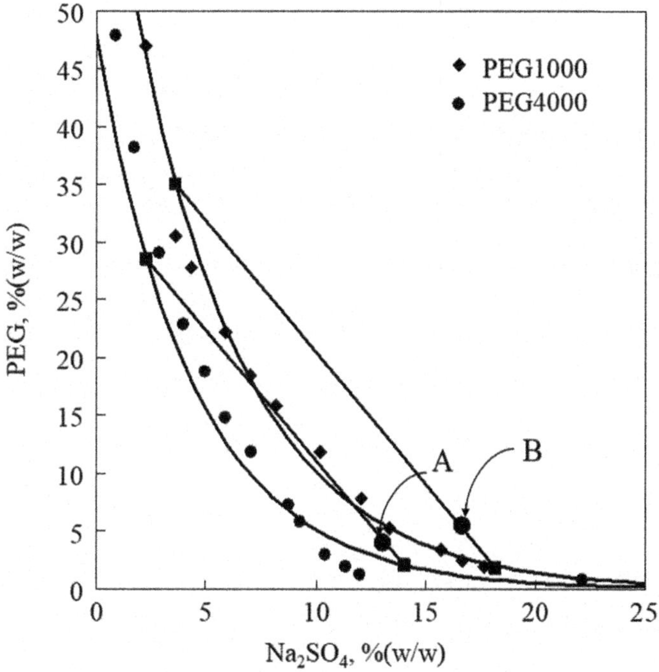

Fig. 5.3 Phase diagrams for PEG/Na_2SO_4 ATPS. Total composition of the ATPS used in the CCC system: **A** 4.0%(w/w) PEG4000/13.0%(w/w) Na_2SO_4, **B** 5.4%(w/w) PEG1000/16.7%(w/w) Na_2SO_4 (reproduced from Ref. [11] by permission of The Japan Society for Analytical Chemistry)

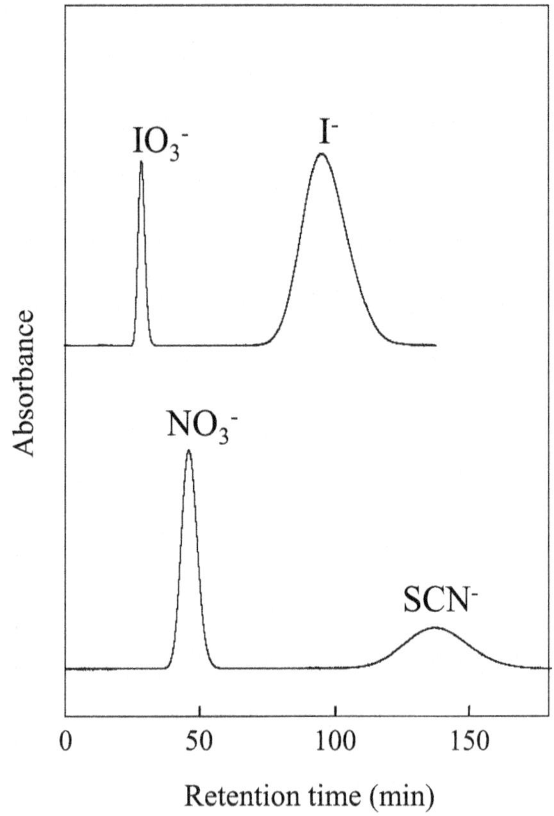

Fig. 5.4 Separation of inorganic anions by a type-J coil planet centrifuge CCC. Total composition of the ATPS: 5.4%(w/w) PEG1000/ 16.7%(w/w) Na_2SO_4 (B in Fig. 5.3); mobile phase: the lower salt-rich phase; flow rate: 2.0 mL min^{-1}; column rotation speed: 1000 rpm; detection wavelength: 220 nm (reproduced from Ref. [11] by permission of The Japan Society for Analytical Chemistry)

where α is the separation factor, N is the number of the theoretical plates, D_2 is the distribution ratio of the compound 2, and V_m and V_s are the volumes of the mobile phase and the stationary phase in the column, respectively [4]. The separation factor is related to the distribution ratio by

$$\alpha = \frac{D_2}{D_1} \tag{5.5}$$

Here the compound 1 elutes faster than the compound 2 ($D_1 < D_2$).

Equation (5.4) indicates that the increase of the R_s value with an increase in the column rotation speed is caused by an increase in the stationary phase volume or the theoretical plate number since the distribution ratio and the separation factor are constant independent of the column rotation speed. Figure 5.6 shows the effect of the column rotation speed on the stationary phase retention ratio, S_f, defined as

$$S_f = \frac{V_s}{V_t} = 1 - \frac{V_m}{V_t} \tag{5.6}$$

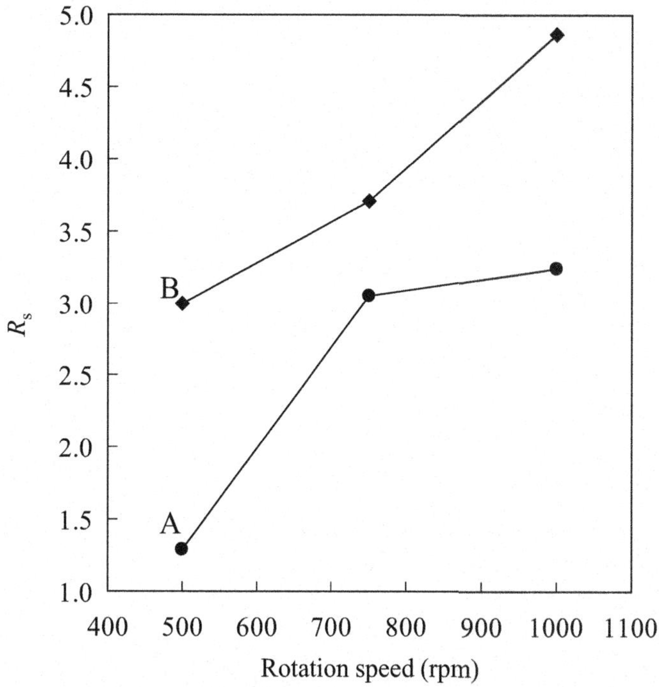

Fig. 5.5 Dependence of the resolution of iodate ion and iodide ion on the column rotation speed. Total composition of the ATPS: **A** 4.0%(w/w) PEG4000/13.0%(w/w) Na_2SO_4, **B** 5.4%(w/w) PEG1000/16.7%(w/w) Na_2SO_4; mobile phase: the lower salt-rich phase; flow rate: **a** 3.0 mL min^{-1}, **b** 3.5 mL min^{-1} (reproduced from Ref. [11] by permission of The Japan Society for Analytical Chemistry)

where V_t is the column volume equal to the sum of V_m and V_s [10]. As observed in Fig. 5.6, the volume of the stationary phase decreases as the column rotation speed and flow rate increase, indicating that the result shown in Fig. 5.5 is not attributed to the change in the stationary phase volume.

The dependence of the theoretical plate number of iodide ion on the column rotation speed is shown in Table 5.1. The plate number drastically increases with an increase in the column rotation speed. The results shown in Fig. 5.6 and Table 5.1 indicate that the stationary phase volume decreases with an increase in the column rotation speed, while the plate number increases, and that the latter effect is greater, thus increasing the resolution. The two phases in the column are subjected to a distribution process of repetitive mixing and settling while the mobile phase is steadily passing through the stationary phase. It can be presumed that as the column rotation speed increases, the number of the partition processes increases and the size of the phase droplets decreases due to more intense agitation, which increases the mass transfer rate between the two phases, thereby increasing the theoretical plate number. The effects of the operating conditions of CCC such as the column rotation

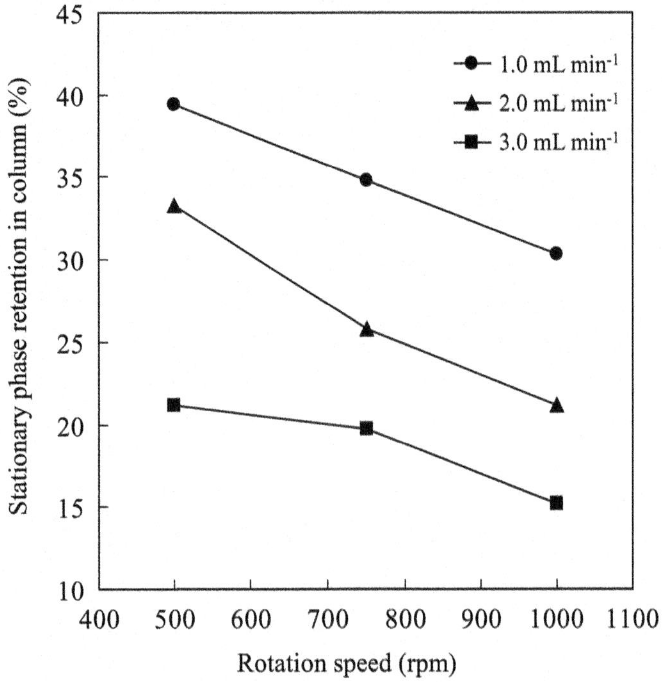

Fig. 5.6 Dependence of the stationary phase volume on the column rotation speed. Total composition of the ATPS: 4.0%(w/w) PEG4000/13.0%(w/w) Na_2SO_4; mobile phase: the lower phase (reproduced from Ref. [11] by permission of The Japan Society for Analytical Chemistry)

speed and the flow rate on the separation efficiency may depend on the composition of the ATPS. The conditions should thus be optimized for each system.

Table 5.1 Theoretical plate number of iodide ion obtained by a type-J coil planet centrifuge CCC (reproduced from Ref. [11] by permission of The Japan Society for Analytical Chemistry)

Composition	Rotation speed (rpm)		
	500	750	1000
A[a]	82	134	193
B[a]	76	137	316

[a]Total composition of ATPS: A 4.0%(w/w) PEG4000/13.0%(w/w) Na_2SO_4; B 5.4%(w/w) PEG1000/16.7%(w/w) Na_2SO_4. Flow rate: A 3.0 mL min^{-1}; B 3.5 mL min^{-1}

5.3 Extraction of Rare Earth Elements with Acetylacetone in PEG/Na_2SO_4 ATPS

Mutual separation of rare earth elements is difficult due to their similar physicochemical properties such as ionic diameter and complexation ability. A lot of solvent extraction techniques have been developed to solve analytical problems and to produce high-purity single rare earth elements [2, 19]. For the separation of rare earth elements by liquid–liquid extraction, organophosphorus extractants such as di-(2-ethylhexyl)phosphoric acid (HDEHP), 2-ethylhexylphosphonic acid mono-2-ethylhexyl ester (HEHEHP), and tributyl phosphate (TBP) are widely used due to their efficient extractability of metal ions into the organic phase [19, 20]. CCC separations of the rare earth elements have thus been carried out using organic solvents such as toluene and *n*-heptane containing some organophosphorus compounds as the stationary phase [21–24]. However, these organophosphorus extracting reagents cannot be used for extraction with ATPS since their solubility in water is very low. Acetylacetone forms stable complexes with rare earth elements, of which the formation constants are substantially large [25], and has moderate hydrophobicity so that not only the reagent itself but also its metal complexes are soluble in the ATPSs [12]. Thus acetylacetone was adopted as the extracting reagent for the study on the enrichment and separation of rare earth elements by multistep pH-peak-focusing CCC.

The distribution behavior of some rare earth elements in PEG/Na_2SO_4 ATPS was investigated using ATPS prepared by mixing 3.00 g of a 50% (w/w) PEG6000, 6.00 g of 20% (w/w) Na_2SO_4, 0.85 g of a buffer solution (1.0 mol L^{-1}), and 0.050 g of acetylacetone. A solution of a rare earth metal ion (1000 ppm, 100 μL) was added to the ATPS prepared and the concentrations of the rare earth metal ion in the upper and lower phases were determined by means of a flow injection analysis system equipped with an absorptiometric detector using 2,7-bis(2-arsonophenylazo)-1,8-dihydroxy-3,6-naphthalenedisulfonic acid (Arsenazo-III) as the chromogenic reagent.

Figure 5.7 shows the plots of log D against pH of the Na_2SO_4-rich phase for La(III), Ce(III), Nd(III), Yb(III), and Sc(III) in the PEG6000/Na_2SO_4 ATPS. All the rare earth metal ions are extracted into the PEG-rich phase as the distribution ratio increases with increasing pH. The order of the distribution ratios of the rare earth metal ions at a given pH is La(III) < Ce(III) < Nd(III) < Yb(III) < Sc(III), which is consistent with the order of formation constants for the complexes with acetylacetone [25, 26]. This result indicates that acetylacetone is a suitable extracting reagent for rare earth elements in ATPS extraction.

The concentrations of sulfate ion and acetylacetone in the present ATPS are 1.5 mol kg^{-1} and 0.01 mol kg^{-1}, respectively, and the acid dissociation constant of acetylacetone is 10^{-9} [25]. On the other hand, the first-order formation constants of sulfate and acetylacetonate complexes of rare earth metal ions, $\log\beta_1$, are 3–4 and 5–8, respectively [25]. Therefore, it may be assumed that rare earth metals exist in the Na_2SO_4-rich phase only as sulfate complexes in acidic region, while they exist as acetylacetonate complexes in the PEG-rich phase, and the distribution ratio of a

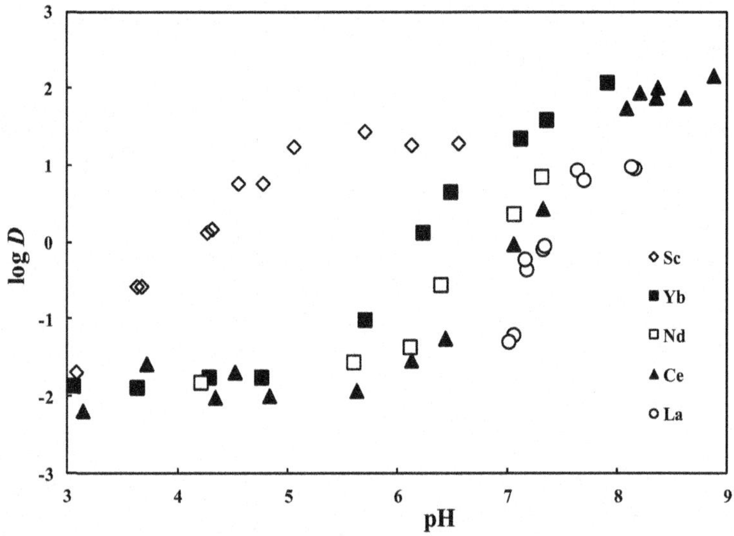

Fig. 5.7 Liquid–liquid extraction of rare earth elements with acetylacetone in PEG6000/Na_2SO_4 ATPS at 25.0 °C. See text for detailed experimental conditions (reproduced from Ref. [12] by permission of The American Chemical Society)

rare earth metal ion, M^{3+}, is given by

$$D = \frac{\sum_n [MA_n^{(3-n)+}]_{PEG}}{\sum_x [M(SO_4)_x^{(3-2x)+}]} = \frac{\sum_n P_{MA_n}\beta_{MA_n}K_a^n[HA]^n}{\sum_x \beta_{M(SO_4)_x}[SO_4^{2-}]^x[H^+]^n} \tag{5.7}$$

where the subscript PEG denotes the PEG-rich phase; HA is acetylacetone; P_{MA_n} is the partition coefficient of the metal acetylacetonate complex; $MA_n^{(3-n)+}$, and β_{MA_n}, and $\beta_{M(SO_4)_n}$ are the overall formation constants of $MA_n^{(3-n)+}$ and $M(SO_4)_x^{(3-2x)+}$ in the Na_2SO_4-rich phase, respectively. The concentrations of free HA in the both phases of the ATPS can be assumed to be constant throughout the experiment because the distribution coefficient of acetylacetone is independent of pH in the range of pH 2–9 and the concentration of a metal ion is negligibly smaller than that of acetylacetone. Since the concentration of sulfate ion in the Na_2SO_4-rich phase can also be assumed to be constant, the following relationship holds between log$D$ and pH from Eq. (5.7):

$$\log D = n\mathrm{pH} + \mathrm{constant} \tag{5.8}$$

Equation (5.8) shows that the slope of the log D vs. pH plot indicates the coordination number of acetylacetonate anions per metal atom in the extracted complex. As shown in Fig. 5.7, the slopes of the log D vs. pH plots for the rare earth metal ions examined range from 1.8 to 2.1 in the linear ranges of the plots, which means that complexes with bimolecular coordination of acetylacetone to metal ions are extracted into the PEG-rich phase. However, the aforementioned assumptions may not hold in the neutral or basic regions, in which case the existence of acetylacetonate complexes in the Na_2SO_4-rich phase is not negligible. In such a case, rare earth metal ions can be extracted into the PEG-rich phase as the neutral tris chelate complexes, MA_3, and the dominant metal complex in the Na_2SO_4-rich phase may be MA^{2+}. Further investigation is needed for elucidation of the structures of the extracted metal species in the PEG/salt ATPS.

5.4 pH Profile of the Column Effluent Formed by pH Stepwise Elution with Buffer Solutions

The results shown in Fig. 5.7 indicate that rare earth metal ions can be separated by varying pH using PEG/Na_2SO_4 ATPS containing acetylacetone as the extracting reagent and the extraction properties of the metal ions can be used to construct optimal chromatographic conditions for multistep pH-peak-focusing CCC for effective separation of the metal ions. For the CCC experiments, PEG1000 was used instead of PEG6000 as a polymer forming ATPS, because the pressure drop along the CCC column with the ATPS formed by PEG6000 is too high to deliver the mobile phase (salt-rich phase) due to high viscosity of the PEG-rich phase. As described in Chap. 2, the distribution coefficients of solute compounds in PEG/salt ATPSs are mainly determined by the difference in concentration of ethylene oxide (EO) units of PEG, ΔEO (mol(EO) kg^{-1}), between the two phases and are independent of molecular weight of PEG [27–31]. The PEG1000/Na_2SO_4 ATPS used for CCC experiments has a ΔEO value of 20 mol kg^{-1}, which is approximately the same as that of the PEG6000/Na_2SO_4 ATPS (23 mol(EO) kg^{-1}) used for the investigation of the distribution behavior of the rare earth metal elements.

The CCC column was a 9.3 m PTFE tube with an inner diameter of 1.5 mm and an internal volume of 16.5 mL. The column rotation speed was set at 750 rpm. The ATPS was prepared by dissolving 108 g of PEG1000, 334 g of Na_2SO_4, and 10 g of acetylacetone in water to make the total mass of 2000 g. The Na_2SO_4-rich phase with lower viscosity was used for the mobile phase, while the PEG-rich phase for the stationary phase. A sodium hydroxide solution was added to the upper PEG-rich phase, with a final pH of 8.3, for trapping all the rare earth metal ions on the top of the CCC column. The column was first filled with the PEG-rich phase, then the mobile phases (Na_2SO_4-rich phase) basified to pH 7.1, 6.5, and 5.9 by adding a buffer reagent were pumped into the column in this order.

Figure 5.8 shows the pH profiles of the column effluents obtained by the pH stepwise gradient elution with the mobile phases containing *N*,*N*-bis(2-hydroxyethyl)-2-aminoethanesulfonic acid (BES), piperazine-1,4-bis (2-ethanesulfonic acid) (PIPES), and 2-(*N*-morpholino) ethanesulfonic acid (MES). The concentration of the buffer reagent in all the mobile phases was set at 0.050 mol L^{-1}. First, a portion

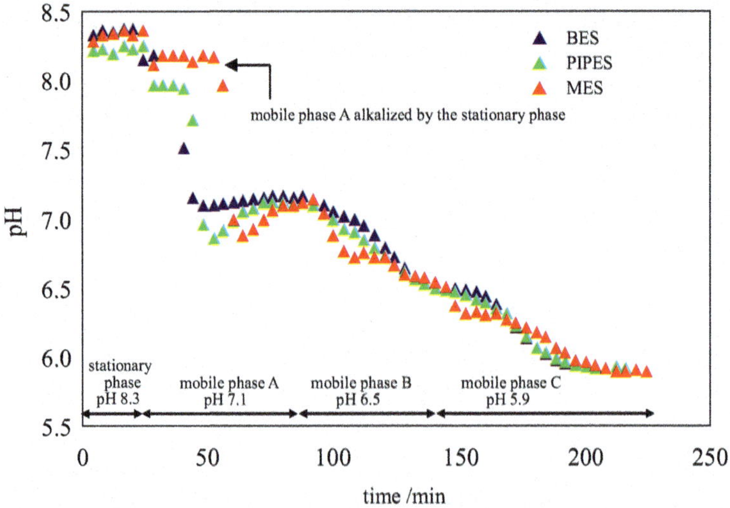

Fig. 5.8 pH profiles of the column effluents obtained by stepwise elution of the mobile phases of different pH adjusted by an identical buffer. Buffer concentration in each mobile phase is 0.050 mol L^{-1}. (reproduced from Ref. [12] by permission of The American Chemical Society)

of the stationary phase was extruded from the column, followed by elution of the first mobile phase (A) alkalized to ca. pH 8.0 by the stationary phase of pH 8.3. Subsequently, the mobile phases of pH 7.1 (A), 6.5 (B), and 5.9 (C) eluted sequentially from the column. However, this pH stepwise gradient elution did not form a clear pH jump between the two mobile phases with different pH, which means that the dispersion of the pH border zones between the mobile phases of different pH prepared with the same buffer reagent is not suppressed by a neutralization reaction.

In addition, as can be seen from Fig. 5.8, the pH of the first mobile phase A containing PIPES or MES decreased down to the value lower than the original pH at the front of the elution zone. The decrease in pH from the optimum value at the pH border must be avoided because coexisting metal ions may co-elute with the target metal ion. This phenomenon is caused by the difference in partition coefficient between the acid and base forms of the buffer reagent in the ATPS and occurs at the boundary between the mobile phases having small buffer capacities.

Taking MES as an example, its distribution ratio is expressed as

$$D_{\mathrm{MES}} = \frac{[\mathrm{HMES}]_{\mathrm{PEG}} + [\mathrm{MES}^-]_{\mathrm{PEG}}}{[\mathrm{HMES}] + [\mathrm{MES}^-]} = \alpha_{\mathrm{acid}} P_{\mathrm{acid}} + \alpha_{\mathrm{base}} P_{\mathrm{base}} \tag{5.9}$$

where α_{acid} and α_{base} are the stoichiometric fractions of the acid form (HMES) and base form (MES^-) of MES, respectively, while P_{acid} and P_{base} are the partition coefficients of HMES and MES^- defined as follows:

$$P_{\mathrm{acid}} = \frac{[\mathrm{HMES}]_{\mathrm{PEG}}}{[\mathrm{HMES}]} \tag{5.10}$$

$$P_{\mathrm{base}} = \frac{[\mathrm{MES}^-]_{\mathrm{PEG}}}{[\mathrm{MES}^-]} \tag{5.11}$$

Figure 5.9 shows the variation of the distribution ratio of MES with pH of the Na_2SO_4-rich phase. The P_{acid} and P_{base} values can be estimated to be 0.45 and 3.18 from the D values of MES under acidic and basic conditions. These results indicate that the base form of MES distributes more into the stationary PEG-rich phase than the acid form, so that as the zone containing MES buffer moves through the column, the concentration of HMES at the front of the zone is relatively higher than that of MES^-, resulting in a lower pH of the mobile Na_2SO_4-rich phase. As the elution proceeds, the pH of the column effluent approaches the value of the original mobile phase containing MES buffer since the stationary phase is continuously equilibrated with the mobile phase. On the contrary, at the rear end of the MES buffer elution zone, the concentration of MES^- becomes higher than that of HMES, making the pH of the mobile phase higher.

Figure 5.10 represents the pH profile of the column effluent obtained by consecutive introduction of the Na_2SO_4-rich phase of pH 5.9 containing 0.050 mol L^{-1} MES buffer (mobile phase A) and that free from a buffer (the mobile phase B) into the column filled with the buffer-free PEG-rich phase. This experimental result shows that the changes in pH at the borders between the mobile phases of different pH are caused by the differential distribution of the acid and base forms of a buffering agent.

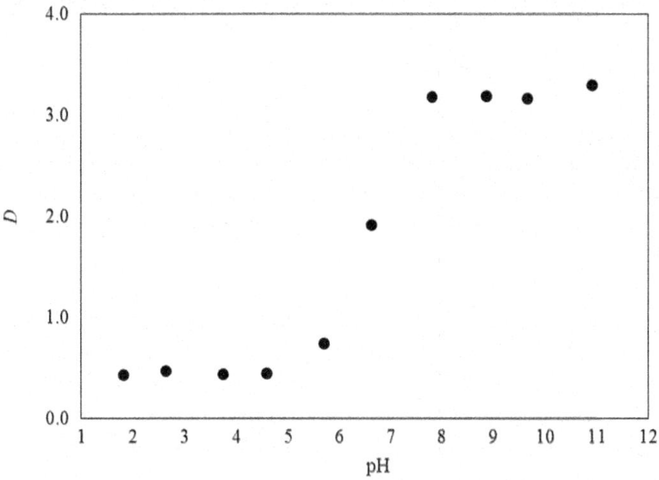

Fig. 5.9 Dependence of the distribution ratio of MES on pH of the lower phase in PEG1000/ Na_2SO_4 ATPS. Composition of the total system: PEG1000 = 5.4% (w/w), Na_2SO_4 = 16.7% (w/ w). Temperature: 25 °C (reproduced from Ref. [12] by permission of The American Chemical Society)

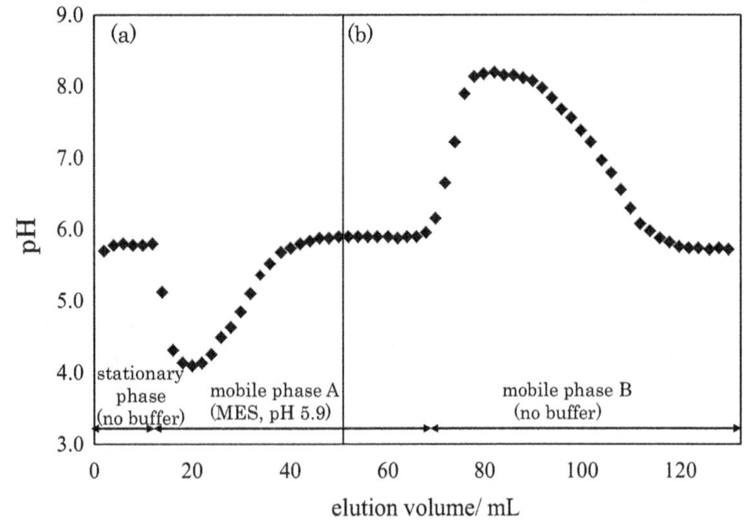

Fig. 5.10 pH profile of the column effluent obtained by stepwise introduction of the mobile phase A of pH 5.9 containing 0.050 mol L^{-1} MES into the column filled with the stationary phase free from a buffer (**a**) and the mobile phase B free from a buffer into the column after the equilibration with the mobile phase A (**b**) (reproduced from Ref. [12] by permission of The American Chemical Society)

5.5 Simultaneous Separation and Enrichment of Rare Earth Elements by Multistep pH-Peak-Focusing CCC

In order to suppress the undesirable pH changes at the border between two mobile phases with different pH, it is necessary to employ the mobile phases with high buffer capacity at the pH. Therefore, the pH of each mobile phase should be adjusted using a buffer reagent that exhibits maximum buffer capacity at the pH. Figure 5.11 shows the pH profile of the column effluent obtained by a stepwise gradient elution with the mobile phases of which the pH values were adjusted with BES (pH 7.1), PIPES (pH 6.5), MES (pH 5.9), and H_2SO_4 (pH 3.0); the pK_a values of BES, PIPES, and MES are 7.15, 6.8, and 6.15, respectively [32]. As shown in Fig. 5.11, this gradient system provided sharp pH jump borders. At the front of the elution zone of the mobile phase D (pH 3.0), the pH of the column effluent is higher than that of the mobile phase C (pH 5.9) as predicted from the result shown in Fig. 5.10b, since the mobile phase D, of which the pH was adjusted with H_2SO_4, has little buffer capacity. However, this change in pH does not adversely affect separation of metal ions. Rather a steeper pH jump may promote the enrichment.

Chromatographic separation of La(III), Nd(III), Yb(III), and Sc(III) obtained by this pH stepwise gradient elution is shown in Fig. 5.11. The sample solution (1 mL) containing 1.25×10^{-5} mol L^{-1} of each rare earth metal ion was prepared with the Na_2SO_4-rich phase to be used as the first mobile phase (mobile phase A) in a

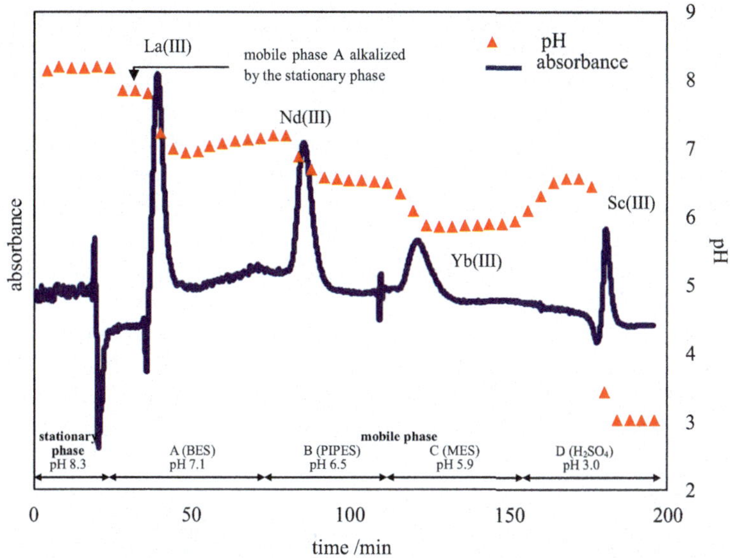

Fig. 5.11 pH profile of the column effluent and a chromatogram for La(III), Nd(III), Yb(III), and Sc(III) obtained by stepwise elution of the mobile phases of which the pH values were adjusted with BES (pH 7.1), PIPES (pH 6.5), MES (pH 5.9), and H_2SO_4 (pH 3.0). Stationary phase: pH 8.3 (5.9 mL); mobile phase: A = pH 7.1 (0.050 mol L^{-1} BES), B = pH 6.5 (0.050 mol L^{-1} PIPES), C = pH 5.9 (0.050 mol L^{-1} MES), D = pH 3.0 (H_2SO_4). Sample solution: 1.25×10^{-5} mol L^{-1} rare earth metal ion (1 mL). Flow rate: 0.50 mL min^{-1}. S_f: 36% (reproduced from Ref. [12] by permission of The American Chemical Society)

stepwise gradient elution and injected into the CCC column filled with the PEG-rich phase of pH 8.3. The mobile phases were then pumped through the column at a flow rate of 0.5 mL min^{-1} while the column was rotated at 750 rpm. The operating pressure was 1.5 MPa and the S_f value ranged from 28 to 36%. The metal ions eluted from the column were detected by measuring the absorbance of colored complexes at 650 nm by a post-column reaction with 0.1 mmol L^{-1} Arsenazo-III aqueous solution acidified to pH 2.0 with HCl. As shown in Fig. 5.11, all four metal ions are completely separated, and the elution peaks are very sharp compared to the chromatographic peaks usually observed in CCC. The retention time of each rare earth metal ion coincides with that of the corresponding pH border, indicating that these very sharp peaks are obtained by pH-peak-focusing.

Ce(III) coeluted with La(III) in this pH stepwise gradient system. Thus, the separation of La(III) and Ce(III) as well as Nd(III), Yb(III), and Sc(III) was examined employing a gradient elution consisting of five pH borders. A higher pH value of the stationary phase (pH 8.5) was adopted and the mobile phase X of which the pH was adjusted by 2-[4-(2-Hydroxyethyl)-1-piperazinyl]ethanesulfonic acid (HEPES, pK_a = 7.55 [32]) to 7.5 was inserted before the mobile phase A of pH 7.1. Figure 5.12 shows a chromatogram obtained by the modified pH stepwise gradient elution system, representing that successful separation of Ce(III) and La(III) can be achieved.

Multistep pH-peak-focusing CCC enables one to perform simultaneous separation and enrichment of metal ions. Figure 5.13 represents a chromatogram of the rare earth metal ions obtained by an injection of a sample solution containing 3.12×10^{-7} mol L^{-1} of each metal with a 40 mL sample loop and the elution with the same pH stepwise gradient system as shown in Fig. 5.12. In order to enrich the metal ions completely at the top of the column, the pH values of the sample solution and the stationary phase were adjusted to pH 9.4 and 9.7, respectively. The concentrations of the buffer reagents added to the mobile phases X-D were accordingly increased to 0.15 mol L^{-1} to neutralize the stationary phase of a higher pH value and suppress undesirable pH change at the pH borders between the mobile phases of different pH values. The peak widths of the metal ions are approximately equal to those obtained when 1 mL sample solution is injected (Fig. 5.12) and are less than the sample volume (4 mL for Sc(III) to 13 mL for La(III)), indicating that the multistep pH-peak-focusing works well even when a sample solution larger than the column volume is injected.

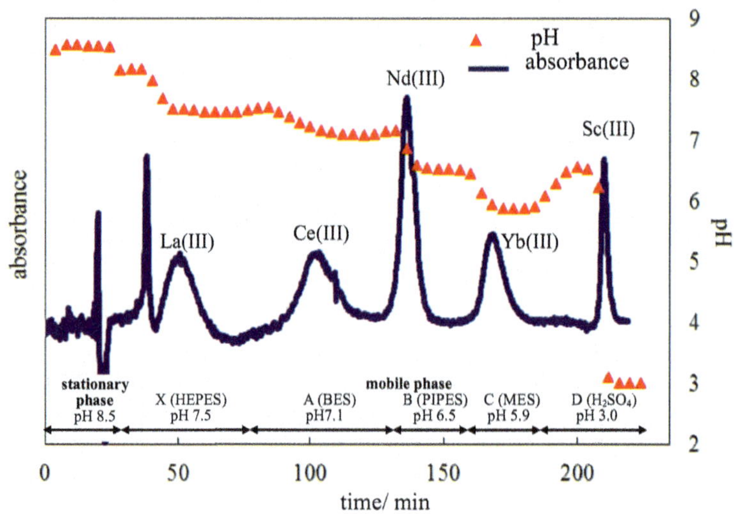

Fig. 5.12 Separation of La(III), Ce(III), Nd(III), Yb(III), and Sc(III) by pH stepwise gradient elution. Stationary phase: pH 8.5 (5.8 mL); mobile phase: X = 7.5 (0.050 mol L^{-1} HEPES), A = pH 7.1 (0.050 mol L^{-1} BES), B = pH 6.5 (0.050 mol L^{-1} PIPES), C = pH 5.9 (0.050 mol L^{-1} MES), D = pH 3.0 (H_2SO_4). Sample solution: 1.25×10^{-5} mol L^{-1} rare earth metal ion (1 mL). Flow rate: 0.50 mL min^{-1}. S_f: 35% (reproduced from Ref. [12] by permission of The American Chemical Society)

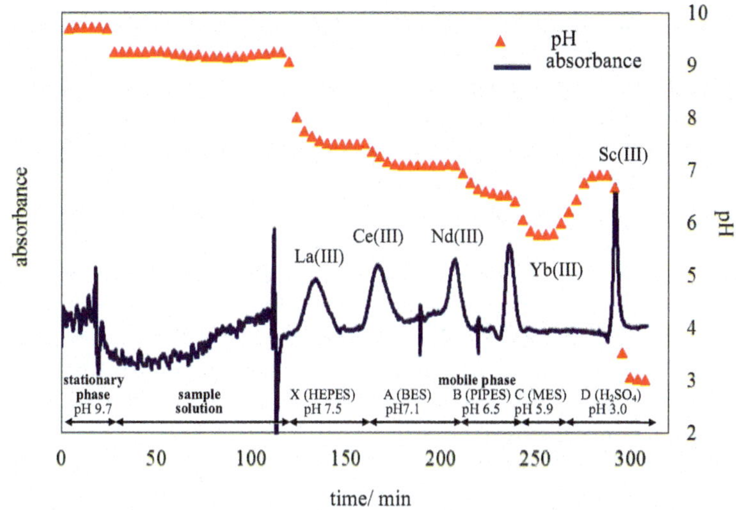

Fig. 5.13 Enrichment separation of La(III), Ce(III), Nd(III), Yb(III), and Sc(III) by pH stepwise gradient elution. Stationary phase: pH 9.7 (4.6 mL); mobile phase: X = 7.5 (0.15 mol L^{-1} HEPES), A = pH 7.1 (0.15 mol L^{-1} BES), B = pH 6.5 (0.15 mol L^{-1} PIPES), C = pH 5.9 (0.15 mol L^{-1} MES), D = pH 3.0 (H_2SO_4). Sample solution: 3.12×10^{-7} mol L^{-1} rare earth metal ion (40 mL). Flow rate: 0.50 mL min^{-1}. S_f: 28% (reproduced from Ref. [12] by permission of The American Chemical Society)

5.6 Conclusions

CCC is an attractive separation method, which enables one to perform efficient chromatographic separation with ATPS without any solid support. One of the two phases of the ATPS is held in a chromatographic tube as the stationary phase by centrifugal force, while the other liquid phase is pumped through it as the mobile phase. An advantage of CCC is that the stationary phase retention ratio, S_f, is higher than that of HPLC [4]. This large S_f value results in larger capacity of a CCC column for the same total column volume and the higher resolution for a given theoretical plate number compared to HPLC. This makes CCC an attractive method for preparative-scale chromatography.

pH-peak-focusing CCC is a unique technique developed for the enrichment of metal ions and acidic and basic compounds on the basis of neutralization during elution between the mobile phase and the stationary phase. The use of ATPS as a separation platform of CCC allows the formation of multistep pH jumps, enabling simultaneous separation as well as enrichment of metal ions.

Multistep pH-peak-focusing CCC with PEG/Na_2SO_4 ATPS was developed and applied to the separation and enrichment of rare earth metal ions. Acetylacetone was adopted as the extracting reagent of the metal ions and the PEG-rich phase and salt-rich phase of the ATPS were used as the stationary phase and mobile phase of CCC, respectively. A sample solution of metal ions was introduced into the CCC column filled with the stationary phase containing acetylacetone alkalized by sodium hydroxide, and then the mobile phases of which the pH values had been adjusted with buffer reagents were pumped into the column by stepwise gradient elution in decreasing pH order. Each metal ion was concentrated at a pH border formed in the CCC column through extraction as acetylacetonate complexes into the stationary phase at the front side of the border and back extraction as free metal ions or sulfate complexes into the mobile phase at the acidic rear side of the border. The buffer capacities of the mobile phase and the stationary phase should be high enough so that the pH jumps are not disturbed by the difference in distribution of the acid and base forms of the buffer in the ATPS. Enrichment and separation of La(III), Ce(III), Nd(III), Yb(III), and Sc(III) were simultaneously achieved by multistep pH-peak-focusing CCC using five-step pH gradient elution. Multistep pH-peak-focusing CCC with PEG/salt ATPS has a great potential as a useful method for enrichment and separation of metal ions.

References

1. Ito Y, Conway WD (eds) (1996) High-speed countercurrent chromatography. John Wiley & Sons, New York
2. Berthod A, Maryutina T, Spivakov B, Shpigun O, Sutherland IA (2009) Countercurrent chromatography in analytical chemistry. Pure Appl Chem 81:355–387
3. Ito Y (2005) Countercurrent chromatography. In: Worsfold P, Townshend A, Poole C (eds) Encyclopedia of analytical science, 2nd edn. Elsevier, Amsterdam, pp 241–252
4. Foucault AP (1991) Countercurrent chromatography. Anal Chem 63:569A-579A
5. Shinomiya K, Kabasawa Y, Yanagidaira K, Sasaki H, Muto M, Okada T, Ito Y (2003) Protein separation by nonsynchronous coil planet centrifuge with aqueous–aqueous polymer phase systems. J Chromatogr A 1005:103–112
6. Shibusawa Y, Kinebuchi T, Shimizu M, Shindo H, Ito Y (2003) Purification of single-strand DNA binding protein from an *Escherichia coli* lysate using counter-current chromatography, partition and precipitation. J Chromatogr B 793:275–279
7. Yanagida A, Isozaki M, Shibusawa Y, Shindo H, Ito Y (2004) Purification of glucosyltransferase from cell-lysate of *Streptococcus mutans* by counter-current chromatography using aqueous polymer two-phase system. J Chromatogr B 805:155–160
8. Ding J, Li S, Zhao Y, Guan YH, Deng L, Deng Q (2017) Properties of hydrodynamic J-type countercurrent chromatography for protein separation using aqueous two-phase systems: with special reference to constructing conical columns. J Chromatogr A 1499:103–110
9. Song G, Du Q (2010) Isolation of a polysaccharide with anticancer activity from *Auricularia polytricha* using high-speed countercurrent chromatography with an aqueous two-phase system. J Chromatogr A 1217:5930–5934
10. Ruiz-Angel MJ, Pino V, Carda-Broch S, Berthod A (2007) Solvent systems for countercurrent chromatography: an aqueous two phase liquid system based on a room temperature ionic liquid. J Chromatogr A 1151:65–73
11. Ishii K, Tanaka Y, Hata K, Goto M, Saitoh K, Minamisawa H, Shibukawa M, (2004) Separation of inorganic ions by high-speed countercurrent chromatography with aqueous biphasic system. Bunseki Kagaku 53:911–917
12. Shimizu K, Kuribayashi H, Watanabe H, Shimasaki T, Azuma K, Horie Y, Saitoh K, Saito S, Shibukawa M (2013) Multistep pH-peak-focusing countercurrent chromatography with a polyethylene glycol-Na_2SO_4 aqueous two phase system for separation and enrichment of rare earth elements. Anal Chem 85:978–984

13. Mokhodoeva O, Rudik I, Shkinev V, Maryutina T (2021) Countercurrent chromatography approach to palladium and platinum separation using aqueous biphasic system. J Chromatogr A 1657:462581
14. Kitazume E, Higashiyama T, Sato N, Kanetomo M, Tajima T, Kobayashi S, Ito Y (1999) On-line microextraction of metal traces for subsequent determination by plasma atomic emission spectrometry using pH peak focusing countercurrent chromatography. Anal Chem 71:5515–5521
15. Kitazume E, Takatsuka T, Sato N, Ito Y (2004) Mutual metal separation system with enrichment using pH-peak focusing countercurrent chromatography. J Liq Chromatogr Relat Technol 27:437–449
16. Ito Y, Shibusawa Y, Fales HM, Cahnmann HJ (1992) Studies on an abnormally sharpened elution peak observed in counter-current chromatography. J Chromatogr 625:177–181
17. Ito Y, Oka H (2009) pH-peak-focusing and pH-zone-refining CCC. In: Cazes J (ed) Encyclopedia of chromatography, 3rd edn. CRC Press, London, pp 1080–1813
18. Wang C, Zuo G, Wang X, Kim HY, Zhao S, Sun W, Tong S, Lim SS (2020) Retention mechanism of pH-peak-focusing in countercurrent chromatographic separation of baicalin and wogonoside from *Scutellaria baicalensis* Georgi. J Sep Sci 43:3806–3815
19. Xie F, Zhang TA, Dreisinger D, Doyle F (2014) A critical review on solvent extraction of rare earths from aqueous solutions. Miner Eng 56:10–28
20. Yang X, Xu L, Zhang A, Xiao C (2023) Organophosphorus extractants: a critical choice for actinides/lanthanides separation in nuclear fuel cycle. Chem Eur J 29:e202300456
21. Kitazume E, Bhatnagar M, Ito Y (1991) Separation of rare earth elements by high-speed countercurrent chromatography. J Chromatogr 538:133–140
22. Nakamura S, Hashimoto H, Akiba K (1997) Enrichment separation of rare earth elements by high-speed countercurrent chromatography in a multilayer coiled column. J Chromatogr A 789:381–387
23. Nakamura S, Hashimoto H, Akiba K, Saito Y (1997) Mutual separation of heavy lanthanoid elements and yttrium by high-speed countercurrent chromatography. Anal Sci 13:525–529
24. Dembowski M, Rowley JE, Boland K, Droessler J, Hathcoat DA, Marchi A, Goff GS, May I (2022) Rare earth element separations by high-speed counter-current chromatography. J Chromatogr A 1682:463528
25. Sillén LG, Martell AE (1964) Stability constants of metal-ion complexes. The Chemical Society, London
26. Suzuki N, Kodera J, Imura H (1987) Extraction equilibrium constants of scandium(III) and yttrium(III) in acetylacetone-organic solvent systems and some comparisons with rare earths(III). Inorg Chim Acta 128:261–266
27. Willauer HD, Huddleston JG, Rogers RD (2002) Solute partitioning in aqueous biphasic systems composed of polyethylene glycol and salt: the partitioning of small neutral organic species. Ind Eng Chem Res 41:1892–1904
28. Willauer HD, Huddleston JG, Rogers RD (2002) Solvent properties of aqueous biphasic systems composed of polyethylene glycol and salt characterized by the free energy of transfer of a methylene group between the phases and by a linear solvation energy relationship. Ind Eng Chem Res 41:2591–2601
29. Shibukawa M, Ichikawa R, Baba T, Sakamoto R, Saito S, Oguma K (2008) Separation selectivity of aqueous polyethylene glycol-based separation systems: DSC, LC and aqueous two-phase extraction studies. Polymer 49:4168–4173
30. Zaslavsky BY, Borovskaya AA, Gulaeva ND, Miheeva LM (1991) Partitioning in aqueous two-phase systems as a method of estimation of the relative hydrophobicity of solutes. J Chem Soc Faraday Trans 87:137–140
31. Eiteman MA, Hassinen C, Veide A (1994) A mathematical model to predict the partitioning of peptides and peptide-modified proteins in aqueous two-phase systems. Biotechnol Prog 10:513–519
32. Good NE, Winget GD, Winter W, Connolly TN, Izawa S, Singh RMM (1966) Hydrogen ion buffers for biological research. Biochem 5:467–477

Chapter 6
Aqueous Liquid Chromatography with Hydrophilic Polymer Gel Columns

Abstract Although polymer-based ATPS represents an eco-compatible alternative to conventional organic/aqueous extraction techniques, it has the disadvantage of the difficulty of separating the target compounds from the phase-forming polymers and recovering the polymers after the extraction. The use of a polymer phase immobilized to a solid support or a cross-linked polymer gel is one of the most effective methods to overcome these problems because the polymer gels or the immobilized polymer matrices are isolated from the aqueous solution phase. This chapter discloses the separation mechanism exerted by some hydrophilic polymer gels conventionally used as column packing materials for aqueous liquid chromatography (dextran, polyacrylamide, polyethylene glycol, and methacrylate) in relation to their hydration structures. Liquid chromatography and differential scanning calorimetry studies indicate that the separation of solute compounds in an aqueous solution with the hydrated cross-linked polymer gels is caused by the partition of the solutes into hydrated polymer gel phase similar to the polymer-based ATPS. Water in the polymer gel phase has a different affinity to solutes from that of the bulk water and functions as a component of the separation media. The stationary phase water shows the melting behavior different from that of the bulk water.

Keywords Polymer-based aqueous two-phase system · Hydrated cross-linked hydrophilic polymer gel · Liquid chromatography · Differential scanning calorimetry · Freezable-bound water · Nonfreezing water

6.1 Introduction

Extraction methods using PEG-based ATPS have attractive advantages of eco-compatibility and low material cost. However, there are several factors that inhibit the widespread use of the ATPS as separation and purification platforms [1, 2]. Due to the high viscosity of the polymer solution, the polymer-based ATPS has the disadvantage of slow phase separation unless centrifugation is used and of not allowing high-performance liquid chromatographic separations with CCC, as mentioned in the

M. Shibukawa, *Polymer-Based Aqueous-Two Phase Systems*,
https://doi.org/10.1007/978-981-96-2888-9_6

preceding chapter. The use of significant amounts of the polymers and/or salts and the difficulty and cumbersomeness of recovering them after the extraction of target compounds may also be important problems of the ATPS extraction even though PEG and the inorganic salts usually used for phase separation do not give serious environmental impacts. Furthermore, if the target compound needs to be isolated or purified, it must be separated from the polymers and salts. Since ATPS contains high concentrations of the phase-forming components, the separation of the target compound from them may not be an easy task.

Among the methods reported so far, the most promising one to overcome these problems may be the use of a polymer phase immobilized to a solid support or a cross-linked polymer gel. Rogers et al. developed a PEG-grafted polystyrene-divinylbenzene copolymer (PS-DVB) resin and have shown that the resin has a similar extraction selectivity for a variety of solute compounds to the PEG-rich phase in PEG/salt ATPS [2–6]. In addition, they found that solutes can partition into the resin at salt concentrations much lower than those required for phase separation in PEG/salt ATPS [3]. These results indicate that the solutes can distribute into the resin phase without adding high concentrations of salt. In addition, as the resin can be used repeatedly by stripping the extracted solutes by elution with water, problems related to polymer recovery are also eliminated. PEG-bonded silica-based columns [7–10] and PEG monolithic capillary columns [11] were also synthesized and have been used for the liquid chromatographic analysis of proteins, natural products, and inorganic ions.

On the other hand, Zeiner et al. studied the sorption behavior of L-serine in the ATPS composed of PEG solution and a cross-linked dextran gel, Sephadex G-50 [12]. It is interesting to note that the concentration and molecular weight of PEG in the solution phase have only small impact on the sorption of L-serine into the Sephadex G-50 gel. This also suggests that the distribution of solutes in ATPS is predominantly caused by partition into preexisting hydrated immobilized polymer phase.

Various hydrophilic polymer gels are used as separation membranes and column packing materials for liquid chromatography (LC) [6, 13–25]. Polymer gels based on PEG or copolymers of PEG, dextrans, polyacrylamides, and methacrylates, etc. are of interest in many areas of research and the separation mechanisms as well as applications of these materials to separation and purification of various compounds have been intensely investigated. This chapter describes the results of studies obtained by LC with hydrophilic polymer gels and differential scanning calorimetry (DSC), focusing on the relationship between the solute separation selectivity of various hydrophilic polymer gels and the hydration structure of the polymer matrices in the gels.

6.2 Liquid Chromatographic Determination of the Amounts of Water Acting as a Component of the Separation Media in Hydrophilic Polymer Gels

Many commercially available hydrophilic polymer column packing materials, such as Sephadex, were originally developed for aqueous size-exclusion chromatography. However, the retention volumes or distribution coefficients of relatively low-molecular-weight compounds usually increase with an increase in their molecular weights on the hydrophilic polymer gel columns, contrary to the results predicted from the size-exclusion mechanism [14–16]. This reveals that the size-exclusion effect is not a predominant factor in the separation of the low-molecular-weight compounds in these systems. Water molecules sorbed in polymer networks should play an important role in separation processes, and therefore an understanding of the states of water in the polymer gel systems is of great importance in the elucidation of the mechanism of separation by hydrophilic polymer gels.

Water sorbed in hydrophilic polymer gels generally exhibits physicochemical properties distinct from those of the bulk water. This means that water sorbed in pores of the polymer matrices may exhibit different affinity to solute compounds from the bulk water, and in LC using a polymer gel column and an aqueous mobile phase, it serves as the stationary phase. Generally, the determination of the mobile phase volume is not easy because no ideal tracer compound is available that explores the mobile phase but does not interact with the stationary phase. However, the volume of the mobile phase or the bulk water phase in an LC column with an aqueous mobile phase, V_{BW}, can be determined using small inorganic ions with the same charge as probes according to the following equation [13–16, 22, 23, 26–29]:

$$V_{\mathrm{BW}} = \frac{V_{\mathrm{A}}^{\mathrm{YX}} V_{\mathrm{B}}^{\mathrm{WZ}} - V_{\mathrm{A}}^{\mathrm{WZ}} V_{\mathrm{B}}^{\mathrm{YX}}}{V_{\mathrm{A}}^{\mathrm{YX}} + V_{\mathrm{B}}^{\mathrm{WZ}} - V_{\mathrm{A}}^{\mathrm{WZ}} - V_{\mathrm{B}}^{\mathrm{YX}}} \quad (6.1)$$

where V_i^{jk} is the retention volume of a probe ion i obtained by elution with the mobile phase solution containing electrolyte jk. This equation is derived from a theoretical consequence that the ratio of the distribution coefficients or the retention factors of two ions with the same charge in a two-phase system is constant regardless of the coexisting background electrolytes, provided that the charge balance is satisfied in either phase and the concentrations of the probe ions, A and B, are negligibly smaller than those of the background electrolytes, YX and WZ [21, 22, 28, 29].

The bulk water phase volumes of the LC columns packed with some commercially available polymer packing materials were calculated from Eq. (6.1) adopting iodate, bromide, nitrate, iodide, and thiocyanate ions as the probe ions and sodium chloride and sodium perchlorate as the background mobile phase electrolytes, of which the concentrations are 0.1 mol L^{-1}. The V_{BW} values obtained for polyacrylamide gels, Bio-Gel P-2 and P-4 (Bio-Rad Laboratories, Richmond, CA, USA), cross-linked dextran gels, Sephadex G-10 and G-15 (Pharmacia Fine Chemicals, Uppsala, Sweden), hydroxylated methacrylic polymer gels, Toyopearl HW-40S, 50S, 55S, and 75S (Tosoh, Tokyo, Japan), and a PEG gel, TSKgel Ether-250 (Tosoh, Tokyo, Japan) are tabulated in Table 6.1 together with the total water volumes in the columns, V_t, and the interstitial or interparticulate volumes, V_{int}. The interstitial volume was assumed to be equal to the retention volume of Blue Dextran 2000 (Pharmacia Fine Chemicals, Uppsala, Sweden; molecular weight = 2×10^6). On the other hand, the V_t value was determined according to the following equation:

$$V_t = \frac{W_t(c) - W_g(c)}{\rho} \tag{6.2}$$

where ρ is the density of water, and $W_t(c)$ and $W_g(c)$ denote the total weight of the contents in the column and that of the dry polymer gel, respectively. As Table 6.1 shows, the order of these values is $V_t > V_{BW} > V_{int}$ for all the columns, indicating that a fraction of water in the polymer gel ($V_{BW} - V_{int}$) is the bulk water, while the other fraction of water ($V_t - V_{BW}$) exhibits different affinity to solute compounds from the bulk water and functions as the stationary phase or a component of the separation media in the gel.

The content of the stationary phase water in each polymer gel, w_X, expressed in g g^{-1} dry gel, can be calculated as follows:

$$w_X = \frac{W_t(c) - W_g(c) - \rho V_{BW}}{W_g(c)} \tag{6.3}$$

On the other hand, the water regain (the weight of water taken up by 1 g dry gel), S_r, is calculated by

$$S_r = \frac{W_t(c) - W_g(c) - \rho V_{int}}{W_g(c)} \tag{6.4}$$

Table 6.1 V_t, V_{BW}, and V_{int} values (mL) for the columns packed with various polymer gel packings

Polymer gel	V_t	V_{BW}	V_{int}
Bio-Gel P-2[a]	3.94	1.97 ± 0.09	1.44
Bio-Gel P-4[a]	4.50	2.87 ± 0.44	1.88
Sephadex G-10[a]	3.88	2.29 ± 0.13	1.90
Sephadex G-15[a]	4.01	2.33 ± 0.05	1.87
Toyopearl HW-40S[a]	3.79	2.01 ± 0.11	1.35
Toyopearl HW-50S[a]	4.20	2.90 ± 0.09	1.43
Toyopearl HW-55S[a]	4.41	3.42 ± 0.06	1.44
Toyopearl HW-75S[b]	12.50	11.62 ± 0.05	5.58
TSKgel Ether-250[c]	1.88	1.43 ± 0.02	0.61

Column dimensions: [a]100 mm × 8 mm I.D., [b]300 mm × 8 mm I.D., [c]150 mm × 4.6 mm I.D

The w_x and S_r values for the hydrophilic polymer gel samples are listed in Table 6.2 [14–16]. Table 6.2 demonstrates that the cross-linked polymer gel which has a smaller degree of cross-linking or a larger exclusion limit has a greater S_r value except for Toyopearl HW-75S; the order of the exclusion limit for polyacrylamide, dextran, and hydroxylated methacrylate gels is Bio-Gel P-2 < P-4, Sephadex G-10 < G-15, and Toyopearl HW-40S < 50S < 55S < 75S, respectively. On the other hand, the dependence of the w_x value on the exclusion limit is not identical for Toyopearl HW, Sephadex G, and Bio-Gel P series. With respect to Toyopearl HW-40S, 50S, and 55S, their w_x values are nearly the same, although their S_r values are quite different. However, for both of Sephadex and Bio-Gel, the packing material which has a larger exclusion limit shows a greater w_x value.

6.3 Thermal Phase Transition Behavior of Water Sorbed in Hydrophilic Polymer Gels

Various experimental methods have been utilized for investigation of the structures and properties of water sorbed in polymer gels: DSC [13–16, 23–41], NMR [35, 42–44], and other techniques [32–36]. Among them, DSC has been widely used to obtain quantitative information on the different states of water in water-swollen polymers. Water incorporated into hydrophilic polymer gels typically exhibits three different DSC behaviors, which can be classified as follows [30]: free water or bulk water; freezable-bound water or intermediate water, which exhibits a melting/crystallization temperature different from bulk water; and nonfreezing water that does not exhibit any detectable solid–liquid phase transition. Freezable-bound water and nonfreezing water are generally considered to be associated with the polymer matrix [13–16, 23–25, 31–33, 38, 39].

Figure 6.1 shows DSC heating curves of water sorbed in Toyopearl HW-40S, 75S, and TSKgel Ether-250 [14]. The samples of Sephadex G-10 and G-15, Bio-Gel P-2 and P-4, and Toyopearl HW-50S and 55S show similar DSC curves to that of HW-40S. As is evident from Fig. 6.1, water sorbed in all the sample polymer gels starts to melt at a temperature lower than that of pure water in the heating process. The water melting at 273 K or over is the bulk water and that melting below 273 K is regarded as the freezable-bound water.

The content of the bulk water, w_{bulk}, and that of freezable-bound water, w_{fb}, expressed in g g^{-1} dry gel, are obtained by Eqs. (2.23) and (2.24) (see Chap. 2), respectively, in a similar manner to the measurement of the values for aqueous PEG solutions. The content of nonfreezing water, w_n, can be calculated by subtracting w_{bulk} and w_{fb} from the total content of water, w_t, as follows:

$$w_n = w_t - w_{bulk} - w_{fb} \tag{6.5}$$

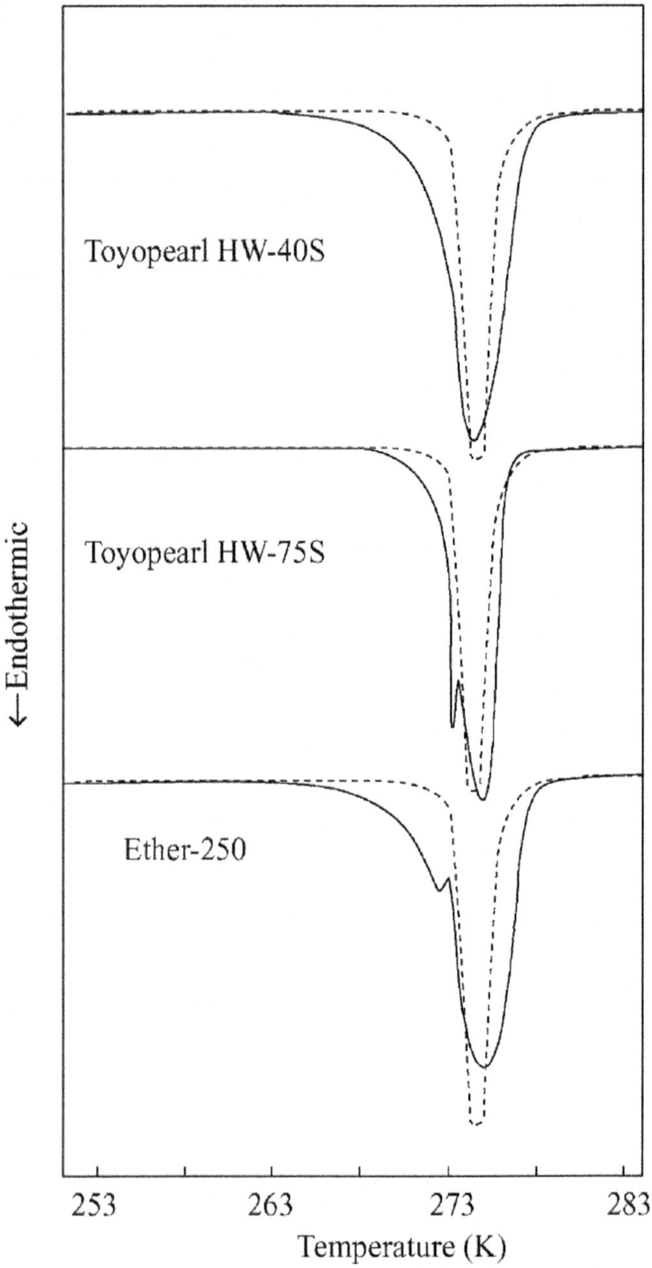

Fig. 6.1 DSC heating curves of water sorbed in TSKgel Toyopearl HW-40S, HW-75S, and TSKgel Ether-250. Broken lines denote the DSC curves for pure water (reproduced from Ref. [14] by permission of Elsevier)

Figure 6.2 shows the variation of the w_{bulk}, w_{fb}, and w_n values with w_t for the sample of Toyopearl HW-40S. The w_{bulk} value decreases as w_t decreases, while both w_{fb} and w_n are constant until the bulk water vanishes, similar to the aqueous solutions of PEG4000 (Fig. 2.17). In the region $w_{bulk} = 0$, w_{fb} decreases linearly with decreasing w_t, while w_n remains constant until the freezable-bound water disappears. The other hydrophilic polymer gels exhibit similar DSC behaviors to that for Toyopearl HW-40S shown in Fig. 6.2.

The w_{fb} and w_n values for each water-swollen polymer gel are summarized in Table 6.3. It should be noted that the sum of w_{fb} and w_n is approximately equal to the w_x value given in Table 6.2 for each polymer gel except Toyopearl HW-75S and TSKgel Ether-250, while the w_x values for these two polymer gels are equal to the w_n values. These results indicate that water interacting differently with ions and molecules in hydrophilic polymer gels from the bulk water exhibits liquid–solid phase transition behavior different from that of the bulk water. For many hydrophilic polymer gels, both nonfreezing water and freezable-bound water function as the

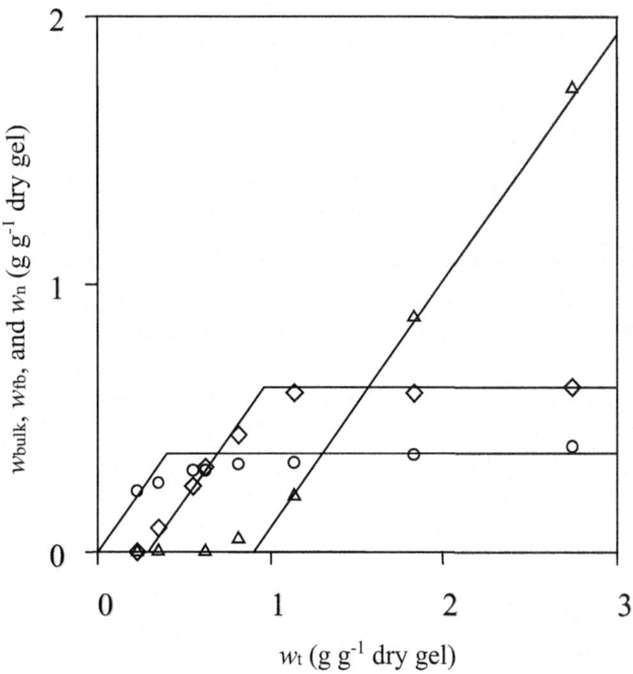

Fig. 6.2 Dependence of w_{bulk}, w_{fb}, and w_n on total water content for Toyopearl HW-40S. Symbols: (△) w_{bulk}; (◇) w_{fb}; (○) w_n (reproduced from Ref. [14] by permission of Elsevier)

separation media similar to aqueous PEG solutions, while for some other gels, only nonfreezing water does.

Contrary to the other hydrophilic polymer gels, a peak of freezable-bound water separated from the bulk water peak was observed for TSKgel Ether-250 and Toyopearl HW-75S as shown in Fig. 6.1. Two clear peaks corresponding to the bulk water and freezable-bound water were also observed for water incorporated in a hydrophobic PS-DVB gel [26] and an octadecyl-bonded (C_{18}) silica gel [27]. However, the repeatability of the peak areas obtained for PS-DVB and C_{18} silica samples is poor; with consecutive measurements, the area of the freezable-bound water peak decreases, while that of the bulk water peak increases. Figure 6.3 shows DSC heating curves of water incorporated in a PS-DVB gel, TSKgel Styrene-250 (Tosoh, Tokyo, Japan), obtained in three consecutive measurements for an identical sample. The results shown in Fig. 6.3 indicate that water detected as the freezable-bound water in the pores of the PS-DVB gel comes out of the pores in the crystallization or melting process and is converted to the bulk water. On the other hand, the reproducible results are obtained for peak areas of the freezable-bound water and the bulk water in the Ether-250 and HW-75S samples. This means that water incorporated in the pores with hydrophobic surface does not strongly interact with the matrices, whereas the hydrated structures of the Ether-250 and HW-75S are stable, even though similar DSC heating curves are observed for water incorporated in the materials.

This interesting result seems to originate from the difference in structures of the polymer gels. Ishikiriyama and Todoki [40] revealed that the mobility of the freezable-bound water observed for poly(methylmethacrylate) hydrogel is similar to that of bulk water from the heat capacity measured by DSC and concluded that the depression of the melting temperature of water in the gel sample is caused by interfacial tension due to isolation of water in the small pores of the gel matrix. Murase et al. [41] also presented a concept of compartmentalized water in small pores in the polymer gel matrices. They noted that water with a melting temperature below 273 K is not necessarily bound water and that compartmentalization of water by the gel network may be responsible for the lower phase transition temperature.

Toyopearl HW gels are packing materials developed for aqueous size-exclusion chromatography. Since Toyopearl HW-75S has the largest exclusion limit (5.00×10^6–11.5×10^6) among the Toyopearl HW gels used [14], the pore size of HW-75S is considered to be much larger than those of the other Toyopearl HW gels; HW-55S has the exclusion limit of 1.05×10^5–1.95×10^5, for example. The S_r value for HW-75S is smaller than that for HW-55S as shown in Table 6.2, probably because Blue Dextran 2000 partially penetrates the pores of the HW-75S gel; the molecular weight of this marker compound is smaller than the exclusion limit of HW-75S.

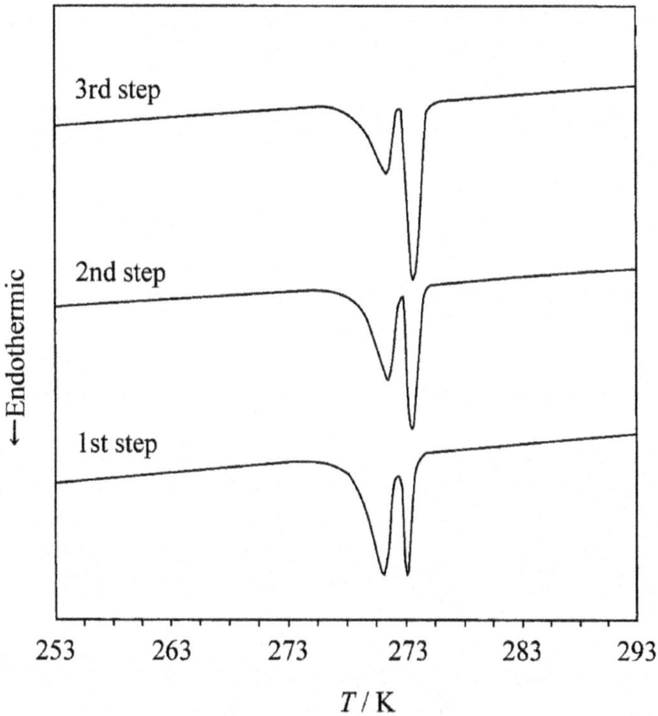

Fig. 6.3 DSC heating curves of water incorporated in TSKgel Styrene-250. $w_t = 0.60$ g g^{-1} dry gel (reproduced from Ref. [26] by permission of Elsevier)

Toyopearl HW-75S and Ether-250 may have heterogeneous structures consisting of polymer-rich domains and water phase located in macropores compartmentalized by the polymer-rich domains. The freezable-bound water observed for these packing materials may be water isolated in the pores of the hydrophobic domains similar to the water incorporated in PS-DVB gels and the depression of the melting point of the freezable-bound water can be ascribed to the isolation of water clusters in the pores [14].

6.4 Solute Retention Selectivity of Hydrophilic Polymer Gel Columns

The retention behavior of various nonionic organic compounds on the columns packed with the water-swollen hydrophilic polymer gels was investigated using pure water as the mobile phase to clarify the relationship between the state of water in the polymer gel phase and the solute retention selectivity [13–16]. The distribution coefficient, K_D, calculated by the following equation was used as the retention parameter:

$$K_D = \frac{V_R - V_{BW}}{V_x} \tag{6.6}$$

where V_R and V_x stand for the retention volume of an organic compound injected into the column and the volume of the stationary phase water in the column, respectively. Although the stationary phase in a column packed with a hydrogel should be regarded as a mixture or solution of the polymer matrix and water, the V_x value was adopted as the stationary phase volume for calculating K_D because it is difficult to evaluate the fraction of the polymer matrix which works as a component of the stationary phase.

In Fig. 6.4, log K_D of alcohols, ketones, nitriles, and nitro compounds obtained on Toyopearl HW-50S, HW-55S, and HW-75S columns are plotted against the values for a Toyopearl HW-40S column. As can be seen from Fig. 6.4, the plots for Toyopearl HW-50S and HW-55S give straight lines with the slopes of approximately unity going through the origin. This indicates that there is little difference in the solute retention selectivity among Toyopearl HW-40S, HW-50S, and HW-55S regardless of the degree of cross-linking or exclusion limit of the polymer gel packing material. In other words, the stationary phases in these three water-swollen Toyopearl HW packings are approximately identical with one another. Figures 6.5 and 6.6 show the plots obtained for the Sephadex G-10 and G-15 and Bio-Gel P-2 and P-4 pairs, respectively. Both of the plots again give straight lines, while the slopes of the plots are not unity but 0.73 and 0.74, respectively. These results mean that the retention selectivities of Sephadex G-15 and Bio-Gel P-4 are lower than those of Sephadex G-10 and Bio-Gel P-2, respectively.

It should be noted that Toyopearl HW-40S, 50S, and 55S have approximately the same w_x values and also exhibit nearly the same solute retention selectivities in contrast to the packing materials of Sephadex G and Bio-Gel P series and Toyopearl HW-75S. For Sephadex G and Bio-Gel P gels, the packing material which has a smaller w_x value shows greater solute retention selectivity when compared with the packing material of the identical series. Toyopearl HW-75S also has the w_x value smaller than those of HW-40S, 50S, and 55S, and shows greater separation selectivity compared with the other materials of Toyopearl HW series. As shown in Chap. 2, the distribution of organic solutes is a function of the degree of phase divergence of polymer-based ATPS as expressed by the difference in polymer concentration between the two phases. The polymer concentration in the polymer gel stationary phase is given by $1/(1 + w_x)$. Since the slope of each plot shown in Figs. 6.4, 6.5, and 6.6 represents the ratio of log K_D obtained for each combination of the polymer gel columns, its value is expected to be equal to the ratio of $1/(1 + w_x)$. The slope of the plot and the ratio of $1/(1 + w_x)$ for each of the polymer gel combinations obtained are compared in Table 6.4. It can be seen that the two values are in fairly good agreement with each other. This suggests that the separation with hydrophilic

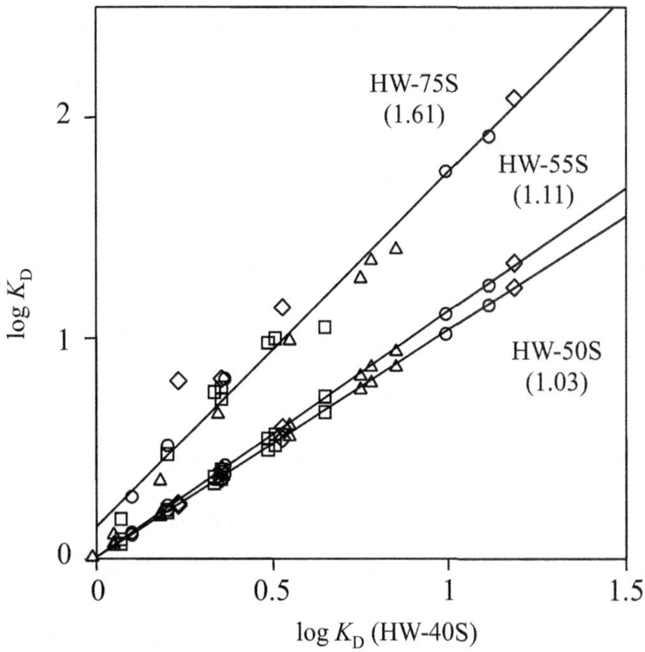

Fig. 6.4 Values of log K_D for Toyopearl HW-50S, 55S, and 75S plotted against log K_D for Toyopearl HW-40S. Values in parenthesis give the slopes of the plots. Symbols: (△) methanol, ethanol, 1-propanol, 1-butanol, 1-hexanol, benzyl alcohol, and phenyl alcohol; (□) acetone, 2-butanone, 3-methyl-2-butanone, 3,3-dimethyl-2-butanone, 2-pentanone, 4-methyl-2-petanone, 3-pentanone, and 2,4-dimethyl-3-pentanone; (○) acetonitrile, butylnitrile, benzonitrile, and phenylacetonitrile; (◇) nitromethane, nitroethane, nitropropane, and nitrobenzene (reproduced from Ref. [14] by permission of Elsevier)

polymer gel materials is caused by the partition of solute compounds into hydrated polymer gel phase similar to the polymer-based ATPS.

When the log K_D values obtained for two different polymer gels are plotted against each other, a single linear relationship is shown only for the same type of polymer gel combination. Figure 6.7 shows the plots of the log K_D values obtained on a Toyopearl HW-40S column against the values on a Bio-Gel P-2 column, for example. Each of the plots for a homologous series of compounds yields a straight line with a similar slope. However, the intercept of the plot depends on the functional group of the compound. Figure 6.4 shows that the log K_D versus log K_D plots for HW-75S are much more scattered compared to the plots for HW-50S and 55S. This can be ascribed to the difference in structure of the polymer gel phase between HW-75S and the other gels of Toyopearl HW series as suggested by the investigation with DSC described in the preceding section.

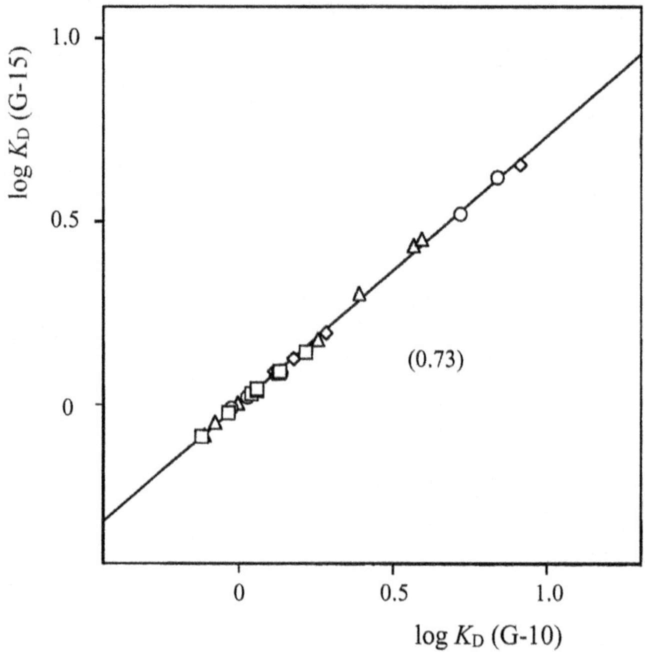

Fig. 6.5 Values of log K_D for Sephadex G-15 plotted against log K_D for Sephadex G-10. See Fig. 6.4 for other details (reproduced from Ref. [15] by permission of Elsevier)

The distribution behavior of some organic compounds in aqueous liquid chromatography with a column packed with TSKgel Ether-250 was compared with that in ATPS formed from PEG and Na_2SO_4 or $(NH_4)_2SO_4$. Figure 6.8 shows the plots of log K_D against the slope of the log D versus ΔEO (mol kg^{-1}) plots at the origin (ΔEO = 0), A, for n-alcohols, ketones, and nitroalkanes; D is defined as Eq. (3.1); and the A value represents the log D at ΔEO of 1 mol kg^{-1} [13]. It should be noted that all the plots yield straight lines with the slope of 21 ± 2 mol (EO) kg^{-1} not only for PEG/Na_2SO_4 but also for PEG/$(NH_4)_2SO_4$ ATPS, while the intercept of the plot is dependent upon hydrophilic functional group of the compound. The slope of the plots represents the ratio of the separation factors of two compounds having different alkyl chain lengths obtained for the PEG gel system to that for the PEG/salt ATPS. The plots shown in Fig. 6.8 indicate that the hydrated TSKgel Ether-250 exhibits the same separation selectivity for methylene groups as that of the aqueous PEG solution with EG concentration of 21 mol kg^{-1}, or PEG weight fraction of 48% (w/w).

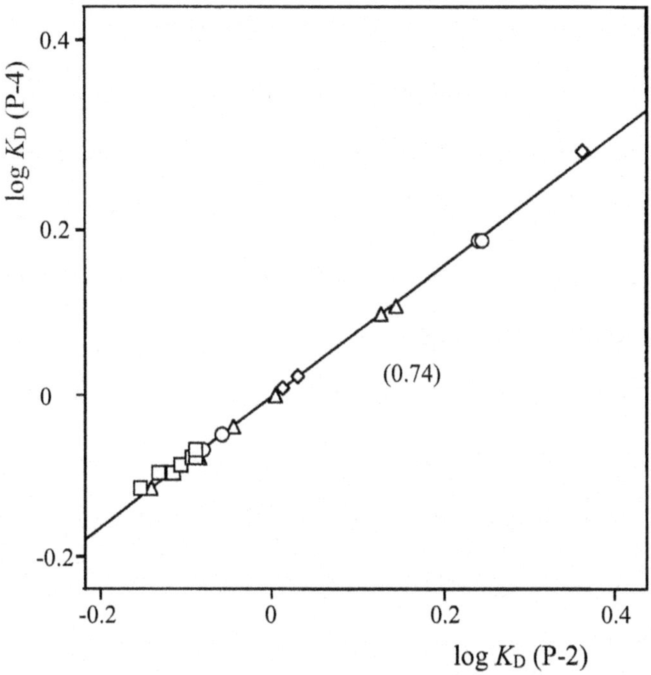

Fig. 6.6 Values of log K_D for Bio-Gel P-4 plotted against log K_D for Bio-Gel P-2. See Fig. 6.4 for other details (reproduced from Ref. [15] by permission of Elsevier)

Table 6.2 w_x and S_r values ($g\,g^{-1}$ dry gel) for water-swollen hydrophilic polymer gel packings

Polymer gel	w_x	S_r
Bio-Gel P-2	1.13 ± 0.05	1.43
Bio-Gel P-4	1.67 ± 0.46	2.70
Sephadex G-10	0.83 ± 0.07	1.04
Sephadex G-15	1.05 ± 0.03	1.34
Toyopearl HW-40S	1.02 ± 0.06	1.40
Toyopearl HW-50S	1.06 ± 0.07	2.28
Toyopearl HW-55S	0.96 ± 0.06	2.93
Toyopearl HW-75S	0.30 ± 0.05	2.59
TSKgel Ether-250	0.49 ± 0.02	1.38

On the other hand, the amount of the stationary phase water in TSKgel Ether-250, w_x, is $0.49 \pm 0.02\ g\,g^{-1}$ dry gel [14], which means that the concentration of polymer matrix in the PEG gel is 67 ± 1% (w/w) provided that the stationary phase can be assumed as a mixture of the polymer matrix and water. Apparently this value does not agree with the value estimated from the slopes of the plots shown in Fig. 6.8 (48% (w/w)). It is plausible that a part of the PEG polymers in the gel phase is not fully

Table 6.3 w_{fb} and w_n values (g g^{-1} dry gel) for water-swollen hydrophilic polymer gel packings

Polymer gel	w_{fb}	w_n
Bio-gel P-2	0.61 ± 0.01	0.64 ± 0.01
Bio-Gel P-4	1.15 ± 0.03	0.52 ± 0.06
Sephadex G-10	0.49 ± 0.02	0.43 ± 0.01
Sephadex G-15	0.71 ± 0.01	0.47 ± 0.02
Toyopearl HW-40S	0.61 ± 0.01	0.40 ± 0.02
Toyopearl HW-50S	0.61 ± 0.02	0.40 ± 0.03
Toyopearl HW-55S	0.56 ± 0.01	0.37 ± 0.00
Toyopearl HW-75S	0.07 ± 0.02	0.31 ± 0.10
TSKgel Ether-250	0.59 ± 0.05	0.44 ± 0.05

Table 6.4 Comparison of the slope of the logK_D versus logK_D plot with the ratio of $1/(1 + w_x)$ for each pair of the polymer gel packings (reproduced from Ref. [16] by permission of The Japan Society for Analytical Chemistry)

	Slope	Ratio of $1/(1 + w_x)$
Sephadex G-15 versus G-10	0.73	0.89
Bio-gel P-4 versus P-2	0.74	0.80
Toyopearl HW-50S versus HW-40S	1.03	0.98
Toyopearl HW-55S versus HW-40S	1.11	1.03
Toyopearl HW-75S versus HW-40S	1.61	1.55

hydrated and may constitute the rigid structure of the gel and the cross-linking agent used for the synthesis of Ether-250 may affect the solute retention. The experimental results obtained by DSC also show that the water-swollen PEG gel and the aqueous PEG solution have different structures from each other (Figs. 2.16 and 6.1).

6.5 Conclusions

The use of a polymer phase immobilized to a solid support or a cross-linked polymer gel is a promising method for avoiding some problems inherent to chemical separation with polymer-based ATPS, such as slow phase separation and cumbersomeness of recovering the phase-forming polymers after the solute extraction. Since the polymer gels are isolated from the solution phase, the problems related to the phase separation and the polymer recovery are eliminated.

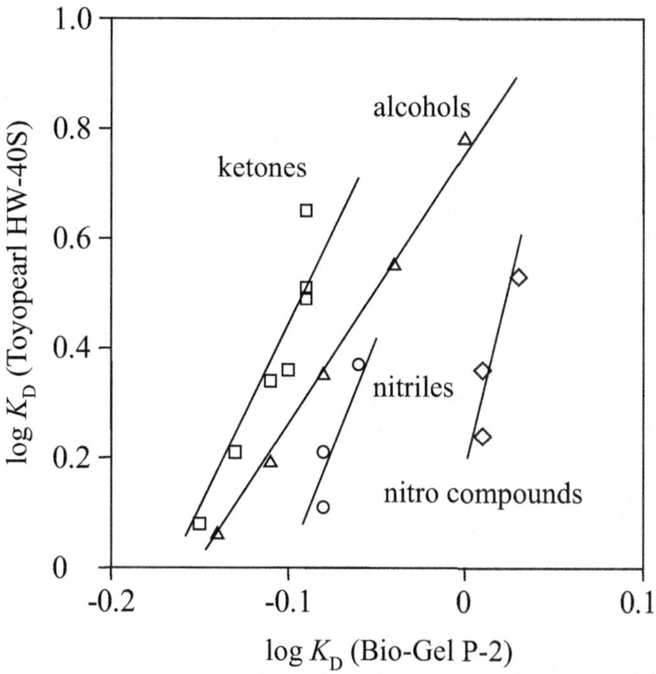

Fig. 6.7 Values of log K_D for Toyopearl HW-40S plotted against log K_D for Bio-Gel P-2. See Fig. 6.4 for other details (reproduced from Ref. [16] by permission of The Japan Society for Analytical Chemistry)

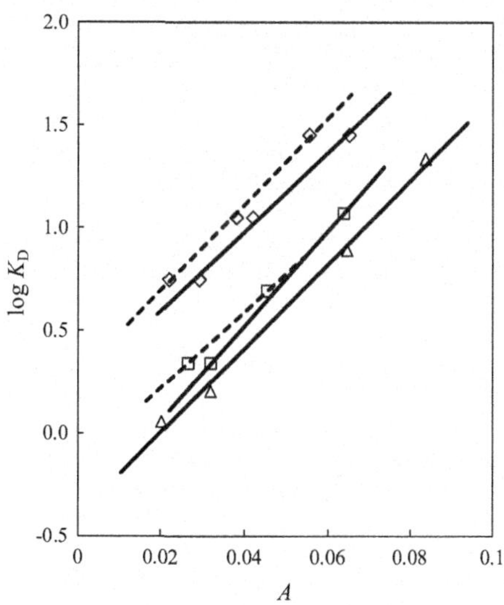

Fig. 6.8 Values of log K_D for TSKgel Ether-250 plotted against A values for PEG/Na_2SO_4 and PEG/$(NH_4)_2SO_4$ ATPS. Symbols: (△) methanol, ethanol, 1-butanol and 1-pentanol; (□) acetone, 2-butanone, and 2-pentanone, 4-methyl-2-petanone, 3–2-pentanone; (◇) nitromethane, nitroethane, and nitropropane. Solid lines and dashed lines denote PEG/Na_2SO_4 and PEG/$(NH_4)_2SO_4$ ATPSs, respectively (reproduced from Ref. [13] by permission of Elsevier)

The separation of solute compounds in an aqueous solution with cross-linked hydrophilic polymer gels is caused by the partition of the solutes into hydrated polymer gel phase similar to the polymer-based ATPS. Water-swollen hydrophilic polymer gels exhibit separation selectivity depending on the structures of the hydrogels and the states of water in the gels. It has been found by the DSC measurements that there are three different states of water in hydrogels, nonfreezing water, freezable-bound water, and bulk water. The evaluation of the amount of stationary phase water by the liquid chromatographic method elucidated that water which has a different affinity to solutes from that of the bulk water and functions as a component of the separation media in the polymer gel phase shows the melting behavior different from that of the bulk water, i.e., nonfreezing water and freezable-bound water.

The distribution of a solute compound is a function of the polymer concentration in the polymer gel stationary phase given by $1/(1 + w_x)$. However, the separation selectivity also depends on the polymer matrix and the structure of the gel phase.

References

1. Karmakar R, Sen K (2019) Aqueous biphasic extraction of metal ions: an alternative technology for metal regeneration. J Mol Liq 273:231–247
2. Dilip M, Griffin ST, Spear SK, Rijksen C, Rodríguez H, Rogers RD (2008) Dual nature of polyethylene glycol-based aqueous biphasic extraction chromatographic (ABEC) resins: uptakes of perchlorate versus mercury(II). Ind Eng Chem Res 47:7390–7396
3. Huddleston JG, Willauer HD, Boaz K, Rogers RD (1998) Separation and recovery of food coloring dyes using ABEC resins. J Chromatogr B 11:237–244
4. Rogers RD, Griffin ST (1998) Partitioning of mercury in aqueous biphasic systems and on ABEC^{TM} resins. J Chromatogr B 711:277–283
5. Griffin ST, Dilip M, Spear SK, Huddleston JG, Rogers RD (2006) The opposite effect of temperature on polyethylene glycol-based aqueous biphasic systems versus aqueous biphasic extraction chromatographic resins. J Chromatogr B 844:23–31
6. Huddleston JG, Looney TK, Broker GA, Griffin ST, Spear SK, Rogers RD (2003) Comparative behavior of poly(ethylene glycol) hydrogels and poly(ethylene glycol) aqueous biphasic systems. Ind Eng Chem Res 42:6088–6095
7. Guo Z, Liu Y, Xua J, Xua Q, Xue X, Zhang F, Ke Y, Liang X, Lei A (2008) Novel reversed-phase high-performance liquid chromatography stationary phase with oligo(ethylene glycol) "click" to silica. J Chromatogr A 1191:78–82
8. Takeuchi T, Oktavia B, Lim LW (2009) Poly(ethylene oxide)-bonded stationary phase for capillary ion chromatography. Anal Bioanal Chem 393:1267–1272
9. Takeuchi T, Lim LW (2010) Multifunctional separation mechanism on poly(oxyethylene) stationary phases in capillary liquid chromatography. Anal Sci 26:937–941
10. Lim LW, Rong L, Takeuchi T (2012) Polyoxyethylene as the stationary phase in ion chromatography. Anal Sci 28:205–213
11. Palm A, Novotony M (1997) Macroporous polyacrylamide/poly(ethylene glycol) matrixes as stationary phases in capillary electrochromatography. Anal Chem 69:4499–4507
12. Mo YL, Fischlschweiger M, Zeiner T (2022) Polymeric aqueous two phase system with an immobilized phase by cross-linking. Ind Eng Chem Res 61:15381–15389
13. Shibukawa M, Ichikawa R, Baba T, Sakamoto R, Saito S, Oguma K (2008) Separation selectivity of aqueous polyethylene glycol-based separation systems: DSC, LC and aqueous two-phase extraction studies. Polymer 49:4168–4173

14. Baba T, Sakamoto R, Shibukawa M, Oguma K (2004) Solute retention and the states of water in polyethylene glycol and poly(vinyl alcohol) gels. J Chromatogr A 1040:45–51
15. Shibukawa M, Aoyagi K, Sakamoto R, Oguma K (1999) Liquid chromatography and differential scanning calorimetry studies on the states of water in hydrophilic polymer gel packings in relation to retention selectivity. J Chromatogr A 832:17–27
16. Shibukawa M (2006) Studies on the states of water in separation media of aqueous liquid chromatography in relation to separation selectivity. Bunseki Kagaku 55:149–162
17. Shibukawa M, Ohta N (1988) Exclusion chromatography of inorganic compounds. In: Dubin PL (ed) Aqueous size-exclusion chromatography. Elsevier, Amsterdam, pp 77–115
18. Hoffman AS (2002) Hydrogels for biomedical applications. Adv Drug Deliv Rev 43:3–12
19. Sassi AP, Shaw AJ, Han SM, Blanch HW, Prausnitz JM (1996) Partitioning of proteins and small biomolecules in temperature and pH sensitive hydrogels. Polymer 37:2151–2164
20. Shibukawa M, Ohta N, Kuroda R (1981) Distribution processes of inorganic solutes in gel chromatography. Anal Chem 53:1620–1627
21. Shibukawa M, Ohta N (1986) Model for partition chromatography of ionic solutes in the presence of eluent electrolytes. Chromatographia 22:261–267
22. Shibukawa M, Ohta N (1988) A new method for the determination of mobile phase volume in normal and reversed-phase liquid chromatography. Chromatographia 25:228–294
23. Shibukawa M, Ohta N, Onda N (1990) Investigation of the states of water in water-swollen hydrogels by liquid chromatography and differential scanning calorimetry. Bull Chem Soc Jpn 63:3490–3494
24. Higuchi A, Iijima T (1985) D.s.c. investigation of the states of water in poly(vinyl alcohol-co-itaconic acid) membranes. Polymer 26:1833–1837
25. Higuchi A, Iijima T (1985) D.s.c. investigation of the states of water in poly(vinyl alcohol) membranes. Polymer 26:1207–1211
26. Baba T, Shibukawa M, Heya T, Abe S, Oguma K (2003) Liquid chromatography and differential scanning calorimetry studies on the states of water in polystyrene-divinylbenzene copolymer gels. J Chromatogr A 1010:177–184
27. Shibukawa M, Kondo Y, Ogiyama Y, Osuga K, Saito S (2011) Interfacial water on hydrophobic surfaces recognized by ions and molecules. Phys Chem Chem Phys 13:15925–15935
28. Shibukawa M, Okutsu H, Saito S (2022) Characterization of the interfacial liquid layer formed on hydrophobic packing material surfaces by liquid chromatographic analysis of the distribution of ions and molecules. ACS Omega 7:15158–15166
29. Shibukawa M (2023) Surface-bubble-modulated liquid chromatography: an experimental strategy for identification of molecular processes of solute retention in reversed-phase separation systems. Anal Sci 39:791–813
30. Hatakeyama T, Quinn FX (1994) Thermal analysis: Fundamentals and applications to polymer science. Wiley, Chichester
31. Ito K, Yoshida K, Ujimoto K, Yamaguchi T (2013) Thermal behavior and structure of low-temperature water confined in Sephadex G15 gel by differential scanning calorimetry and X-ray diffraction method. Anal Sci 29:353–359
32. Ito K, Yoshida K, Bellissent-Funel M-C, Yamaguchi T (2014) Dynamic properties of water confined in Sephadex G15 gel by quasi-elastic neutron scattering and neutron spin echo measurements. Bull Chem Soc Jpn 87:603–608
33. Ping ZH, Nguyen QT, Chen SM, Zhou JQ, Ding YD (2001) States of water in different hydrophilic polymers-DSC and FTIR studies. Polymer 42:8461–8467
34. Qu X, Wirsén A, Albertsson A-C (2000) Novel pH-sensitive chitosan hydrogels: swelling behavior and states of water. Polymer 41:4589–4598
35. Liu WG, Yao KD (2001) What causes the unfrozen water in polymers: hydrogen bonds between water and polymer chains? Polymer 42:3943–3947
36. Iijima M, Hatakeyama T, Hatakeyama H (2021) DSC and TMA studies of polysaccharide physical hydrogels. Anal Sci 37:211–219
37. Bhattacharya S, Shunmugam R (2020) Unraveling the effect of PEG chain length on the physical properties and toxicant removal capacities of cross-linked network synthesized by thiol-norbornene photoclick chemistry. ACS Omega 5:2800–2810

38. Rockwell PN, Maneval JE, Vogel BM, Jablonski EL (2023) Water diffusion and uptake in injectable ETTMP/PEGDA hydrogels. J Phys Chem B 127:5055–5061
39. Vondrasek B, Wen C, Cheng S, Riffle JS, Lesko JJ (2021) On the nature of freezing/melting water in ionic polysulfones. Macromolecules 54:6477–6488
40. Ishikiriyama K, Todoki M (1995) Heat capacity of water in poly(methyl methacrylate) hydrogel membrane for an artificial kidney. J Polym Sci B Polym Phys 33:791–800
41. Murase N, Gonda K, Watanabe T (1986) Unfrozen compartmentalized water in gels and its anomalous crystallization during warming. J Phys Chem 90:5420–5426
42. Kanekiyo M, Kobayashi M, Ando I, Kurosu H, Ishii T, Amiya S (1998) A structural and dynamic study of poly(vinyl alcohol) in the gel state by solid-state ^{13}C NMR and ^{1}H pulse NMR. J Mol Struct 447:49–59
43. Kobayashi M, Ando I, Ishii T, Amiya S (1998) Structural and dynamical studies of poly(vinyl alcohol) gels by high-resolution solid-state ^{13}C NMR spectroscopy. J Mol Struct 440:155–164
44. Machida Y, Kuroki S, Kanekiyo M, Kobayashi M, Ando I, Amiya S (2000) A structural study of water in a poly(vinyl alcohol) gel by ^{17}O NMR spectroscopy. J Mol Struct 554:81–90

Chapter 7
ATPS Chromatographic Techniques for Separation and Enrichment of Metal Ions and Inorganic Anions: Multistep pH-Peak-Focusing Liquid Chromatography and Partition/Ion-Exclusion Chromatographic Ion Stacking

Abstract Liquid chromatography (LC) with a hydrophilic polymer gel column and an aqueous mobile phase is a high-performance separation method based on solute partitioning in a polymer-based ATPS. LC with ATPS, ATPS-LC, provides unique methods for separation and/or enrichment of ionic solutes because both phases of ATPS can dissolve them in a charged state without any special derivatization. This chapter describes the principles and practical applications of two ATPS-LC techniques developed for the separation of metal ions, multistep pH-peak-focusing LC, and enrichment of anionic solutes, partition/ion-exclusion chromatographic ion stacking (PIEC ion stacking). It will be demonstrated that multistep pH-peak-focusing LC is useful for separation and enrichment of rare earth metal ions using the mobile phase containing suitable extractants. Each metal ion is concentrated at the respective pH border in the LC column formed by a stepwise pH gradient elution in a manner similar to multistep pH-peak-focusing CCC. However, the metal enrichment and separation performance of LC with a hydrophilic polymer gel column is much better than that of CCC. PIEC ion stacking is a method that enables one to highly concentrate ionic solutes by the synergism of the effect of background co-ion on the partition of target ions into the polymer gel phase and ion-exclusion effect from the gel phase. A novel analytical methodology for simple and efficient online preconcentration of trace inorganic anions in reagent-grade salt samples for ion chromatographic analysis will be proposed using PIEC ion stacking with a hydrophilic polymer gel column containing fixed anionic charges.

Keywords Aqueous two-phase system · Multistep pH-peak-focusing liquid chromatography · Partition/ion-exclusion chromatographic ion stacking · Hydrated cross-linked hydrophilic polymer gel

M. Shibukawa, *Polymer-Based Aqueous-Two Phase Systems*,
https://doi.org/10.1007/978-981-96-2888-9_7

7.1 Introduction

Liquid chromatography using a column packed with a hydrophilic polymer gel and an aqueous mobile phase is a useful method for high-performance separation based on solute partitioning in ATPS as described in the preceding chapters. Liquid chromatographic techniques also produce useful separations that cannot be achieved by batch extraction methods. Countercurrent chromatography (CCC) is a convenient chromatographic technique that can easily incorporate ATPS, but has drawbacks such as low column efficiency and inevitable loss of the polymer that forms the ATPS. On the contrary, aqueous liquid chromatography with a hydrophilic polymer gel column has a high separation efficiency comparable to that of conventional high-performance liquid chromatography (HPLC) and can achieve separations based on ATPS partitioning without loss of polymers.

Liquid chromatography based on ATPS using hydrophilic polymer gels, ATPS-LC, provides unique methods for the separation and/or enrichment of ionic compounds because the polymer gel phase functioning as the stationary phase can accommodate them without any derivatization. In this chapter, the principles and practical applications of two ATPS-LC techniques developed for the separation and enrichment of metal ions and inorganic anions will be described. The first ATPS-LC technique is multistep pH-peak-focusing liquid chromatography, which shows dramatically improved efficiency for separation of metal ions compared with multistep pH-peak-focusing CCC [1]. The second one is partition/ion-exclusion chromatographic ion stacking, that is, a novel ion enrichment method developed based on the distribution mechanism of ions in ATPS-LC [2, 3]. These two ATPS-LC techniques have great potential as environmentally benign separation and concentration methods for the analyses of metal ions and inorganic anions.

7.2 Multistep pH-Peak-Focusing Liquid Chromatography

7.2.1 *Retention Behavior of Rare Earth Elements on a Hydrophilic Polymer Gel Column with Aqueous Mobile Phases Containing Acetylacetone*

As described in Chap. 5, multistep pH-peak-focusing CCC with polymer/salt ATPS is a useful method that allows simultaneously separate and enrich metal ions [4]. The ATPS is an ideal platform for multistep pH-peak-focusing, because it can meet one of the most important requirements for performing multistep pH-peak-focusing, namely, that both the stationary and mobile phases must have sufficient capacity for both acidic and basic forms of pH buffering agents. However, it is difficult to carry out high speed and highly efficient separation by CCC with the polymer/salt ATPS as the ATPS has high viscosity. The sizes of the phase droplets of the ATPS formed by the agitation during elution through a CCC column are presumed to be relatively large, which reduces the mass transfer rate between the two phases and, consequently, the number of the theoretical plates. In addition, it is not possible to flow a highly viscous mobile phase at high speed in CCC because of the low pressure resistance of the PTFE tubing used as the chromatographic tube. In contrast to CCC, ATPS-LC can use a polymer-free salt solution as the mobile phase, allowing aqueous mobile phase to be pumped at high flow rates. Furthermore, the column efficiency of ATPS-LC is generally much higher than that of CCC since a polymer gel phase with a small particle size can be used as the stationary phase.

As shown in Chap. 6, Toyopearl HW-40S (particle size = 20–40 μm), a cross-linked hydroxylated methacrylic polymer gel, serves as the stationary phase for ATPS-LC when an aqueous mobile phase is used [5–7]. A stainless steel column (150 × 10 mm i.d.) packed with Toyopearl HW-40S was thus adopted for investigating the performance of multistep pH-peak-focusing LC for separation and enrichment of rare earth elements in aqueous solutions. Aqueous sodium sulfate solutions and acetylacetone (acac) were used as the mobile phase and the extractant of rare earth metal ions, respectively. The metal ions eluted from the column were detected by post-column reaction by complexation with Arsenazo-III.

The effect of the mobile phase pH on the retention volumes, V_R, of the rare earth metal ions was investigated using the mobile phases of which the pH was adjusted with a Good's buffer or acetate buffer. Figure 7.1 shows the dependence of the retention volumes of the rare earth metal ions on pH of the mobile phase containing 0.7 mol L^{-1} Na_2SO_4 [1]. All the rare earth metal ions are retained on the Toyopearl HW-40S column in the pH range above 6 by complexation with acac, similar to the retention in CCC using PEG/Na_2SO_4 ATPS [4], and the retention of the rare earths at a given pH increases in the order of the formation constant [8, 9]. This indicates that the Toyopearl HW-40S polymer gel serves as an extraction medium for metal acetylacetonate complexes, as does the PEG-rich phase in PEG/Na_2SO_4 ATPS.

The concentration of the salt in the mobile phase affects the retention of the metal ions. Figure 7.2 shows that the retention volumes of the metal ions decrease with a decrease in Na_2SO_4 concentration in the mobile phase. This result may be partly attributed to the strong salting-out effect of the kosmotropic ions, Na^+ and SO_4^{2-}, [10–13] on the swelling of the gel particles. It has been reported that Na_2SO_4 reduces the osmotic pressure of a hydrophilic polymer gel in water [11], which means that the increase in the salt concentration reduces the swelling of hydrophilic polymer gels, leading to higher polymer concentration in the water-swollen polymer gel phase. The packing material was slurry packed into the column with Na_2SO_4 solution of which the salt concentration is the same as that of the mobile phase to be used. Therefore, the increase in the salt concentration in the mobile phase will increase the amount of the polymer matrix in the column and enhance the extractability of the metal acetylacetonates into the polymer gels. The extraction of metal acetylacetonate complexes into the gel phase may also be facilitated by the salting-out effect of Na_2SO_4.

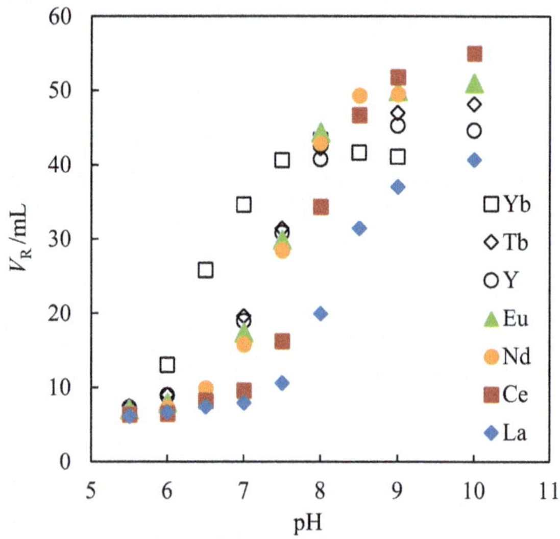

Fig. 7.1 pH dependence of retention volumes of La(III), Ce(III), Nd(III), Eu(III), Y(III), Tb(III), and Yb(III) on a Toyopearl HW-40S column in the presence of acac at 30 °C. Column size: 10 mm i.d. × 150 mm; mobile phase: aqueous solution containing 0.7 mol L^{-1} Na_2SO_4, 50 mmol L^{-1} acac, and 50 mmol L^{-1} pH buffer reagent; sample solution: 1 ppm rare earth metal ions (1.0 mL) (reproduced from Ref. [1] by permission of Elsevier)

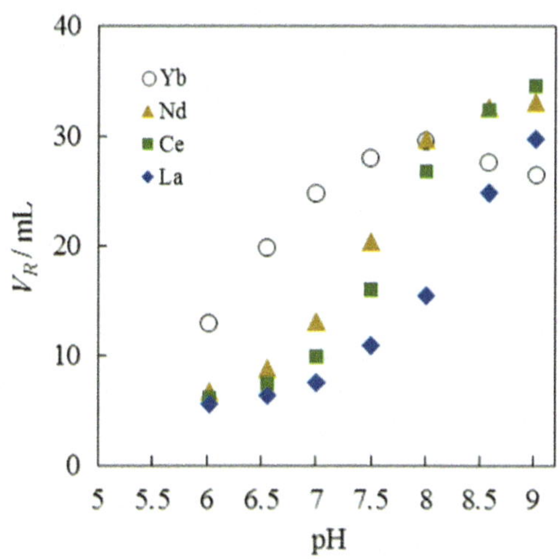

Fig. 7.2 pH dependence of retention volumes of La(III), Ce(III), Nd(III), and Yb(III) on a Toyopearl HW-40S column with the mobile phase containing 0.5 mol L^{-1} Na_2SO_4, 50 mmol L^{-1} acac, and 50 mmol L^{-1} pH buffer reagent. See Fig. 7.1 for other chromatographic conditions (reproduced from Ref. [1] by permission of Elsevier)

7.2.2 *Effect of pH-Peak-Focusing on the Elution Peak of a Rare Earth Element in ATPS-LC*

The optimum pH values of the mobile phases producing pH borders suitable for the enrichment of metal ions by pH-peak-focusing can be obtained from their retention characteristics on the column represented by Fig. 7.1 or Fig. 7.2. Figure 7.3 shows the chromatogram of Nd(III) obtained by pH-peak-focusing with a single-step pH gradient elution with the mobile phases of pH 7.5 and pH 6.5 together with the pH profile of the column effluent; the mobile phases used contain 0.7 mol L^{-1} Na_2SO_4

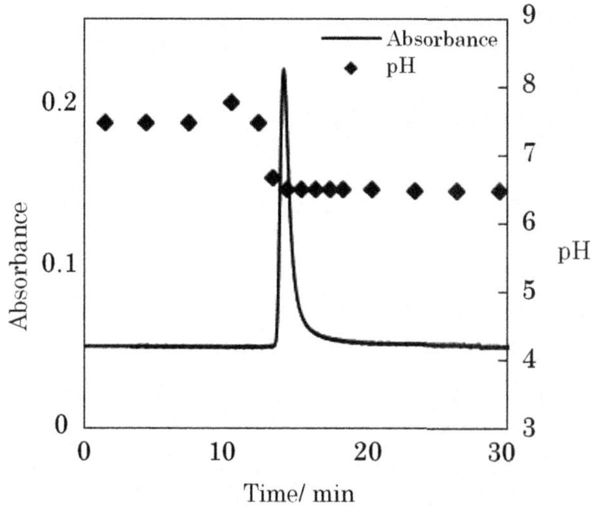

Fig. 7.3 pH profile of the column effluent and a chromatogram for Nd(III) obtained by stepwise elution of the mobile phases of which the pH values were adjusted with HEPES (pH 7.5) and MES (pH 6.5). Flow rate: 1.0 mL min^{-1}; temperature: 30 °C. See Fig. 7.1 for other chromatographic conditions (reproduced from Ref. [1] by permission of Elsevier)

and 50 mmol L^{-1} acac and the pHs of the mobile phases are adjusted with HEPES (pH 7.5) and MES (pH 6.5). Figure 7.4 shows the chromatograms of Nd(III) obtained by isocratic elution with the mobile phases of pHs 6.5 and 7.5 for comparison. It can be seen that the chromatographic peak of Nd(III) obtained by pH-peak-focusing LC (Fig. 7.3) is substantially narrower than the peaks obtained by isocratic elution and the elution time of Nd(III) in pH-peak-focusing ATPS-LC corresponds to that of the pH border. This indicates that the elution peak of a metal ion in ATPS-LC can be effectively sharpened by pH-peak-focusing.

7.2.3 Synergistic Effect of 1,10-Phenanthroline on the Retention of Rare Earth Elements with Acetylacetone

The retention of rare earth metal ions on the Toyopearl HW-40S column with the mobile phase containing only acac as the extraction reagent is not large enough even above pH 9. In addition, some rare earth metal ions precipitate as hydroxides in alkaline solutions above pH 9 [4]. These problems must be overcome to trap the metal ions on the top of the column in a narrow zone for subsequent separation by multistep pH-peak-focusing LC with pH step gradient elution. 1,10-phenanthroline (phen) has been reported to exhibit a remarkable synergistic effect in the extraction of rare earth metal ions into organic solvents with β-diketones due to the formation

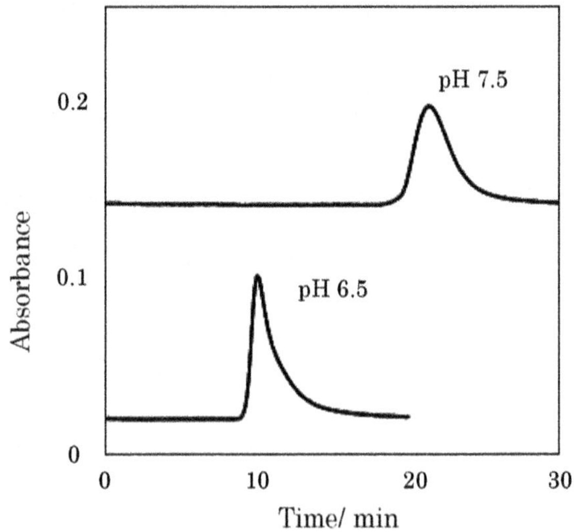

Fig. 7.4 Chromatograms for Nd(III) obtained by isocratic elution of the mobile phases of pH 7.5 and pH 6.5. See Figs. 7.1 and 7.3 for other chromatographic conditions (reproduced from Ref. [1] by permission of Elsevier)

of the ternary complex [14–18]. Therefore, an attempt was made to improve the retention of rare earth metal ions on the polymer column in ATPS-LC by exploiting the synergistic effect of acac and phen.

Figure 7.5 shows the chromatograms of the rare earth metal ions eluted with the mobile phase of pH 5.5 containing 50 mmol L^{-1} acac and 5 mmol L^{-1} phen. The retention volumes of all the metal ions remarkably increased by the addition of phen to the mobile phase. The increase in the retention of the rare earth metal ions on the polymer gel column may be ascribed to the synergistic effect of phen on the extraction of the metal ions with acac into the gel stationary phase by the ternary complex formation in a similar manner to the organic/aqueous two-phase extractions.

On the other hand, the addition of phen to the mobile phase significantly distorted the elution peaks of some rare earth metal ions (Fig. 7.5). The formation of multiple metal complex species containing acac and phen and their slow interconversion reactions may be responsible for this result. The column temperature was therefore increased to accelerate the interconversion reactions. The chromatograms obtained at 50 °C are shown in Fig. 7.6. It can be seen from Fig. 7.6 that increasing the column temperature significantly improved the peak shape of each metal ion.

7.2.4 Multistep pH-Peak-Focusing Separation of Rare Earth Elements by ATPS-LC

Figure 7.7 shows the retention of the rare earth metal ions as a function of the pH of the mobile phase containing 50 mmol L^{-1} acac, 5 mmol L^{-1} phen, and 0.7 mol L^{-1} Na_2SO_4 at 50 °C. As the pH of the mobile phase increases, the retention of each rare

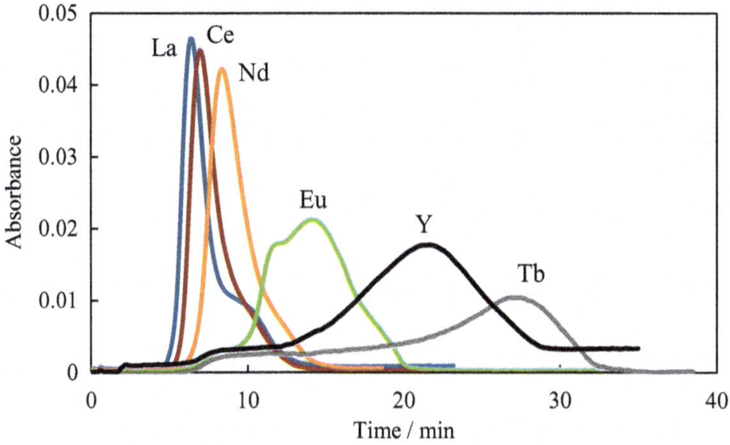

Fig. 7.5 Chromatograms for La(III), Ce(III), Nd(III), Eu(III), Y(III), and Tb(III) obtained by elution with the mobile phase containing 5 mmol L^{-1} phen and 50 mmol L^{-1} acac at 30 °C. Column: Toyopearl HW-40S (10 mm i.d. × 150 mm); mobile phase: aqueous solution containing 0.7 mol L^{-1} Na_2SO_4, 5 mmol L^{-1} phen, 50 mmol L^{-1} acac, and 50 mM MES (pH 5.5); flow rate: 1.0 mL min^{-1}; sample solution: 1 ppm rare earth metal ion (1.0 mL) (reproduced from Ref. [1] by permission of Elsevier)

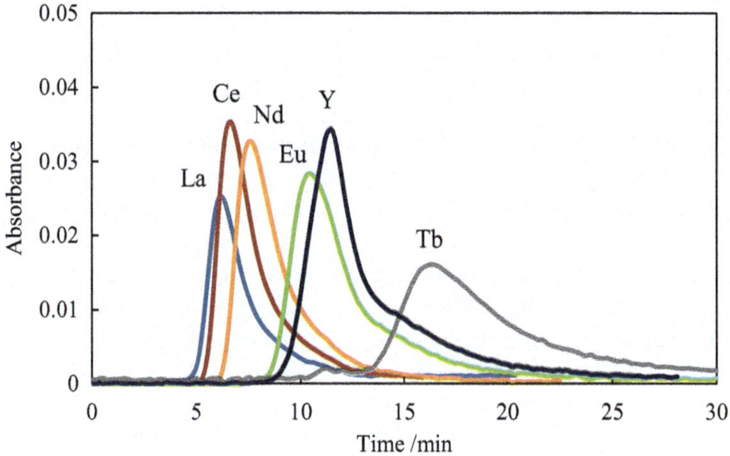

Fig. 7.6 Chromatograms for La(III), Ce(III), Nd(III), Eu(III), Y(III), and Tb(III) obtained by elution with the mobile phase containing 5 mmol L^{-1} phen and 50 mmol L^{-1} acac at 50 °C. See Fig. 7.5 for other chromatographic conditions (reproduced from Ref. [1] by permission of Elsevier)

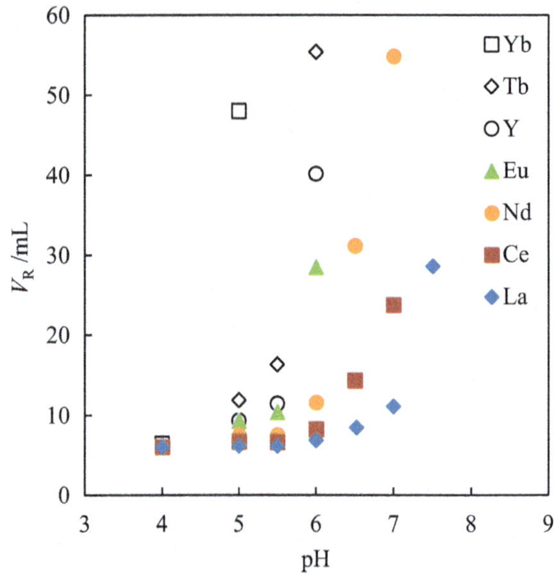

Fig. 7.7 pH dependence of retention volumes of rare earth metal ions on a Toyopearl HW-40S column in the presence of acac and phen at 50 °C. Mobile phase: aqueous solution containing 0.7 mol L^{-1} Na_2SO_4, 5 mmol L^{-1} phen, 50 mmol L^{-1} acac, and 50 mmol L^{-1} pH buffer reagent. See Fig. 7.5 for other chromatographic conditions (reproduced from Ref. [1] by permission of Elsevier)

earth metal ion increases steeply, with all the metal ions being strongly retained in the column at pH 7.5, but at this pH they do not precipitate as hydroxides.

Based on these results, the separation of La(III), Ce(III), Nd(III), Eu(III), Y(III), Tb(III), and Yb(III) was examined by multistep pH-peak-focusing LC. Figure 7.8 shows the chromatogram of the rare earth metal ions and the pH profile of the effluent obtained when the mobile phases containing HEPES (pH 7.5), BES (pH 7.0), MES (pH 6.5, 6.0, and 5.5), and acetic acid buffer (pH 4.0) were passed through the column in this order. La(III), Ce(III), Nd(III), Tb(III), and Yb(III) are eluted from the column with retention times consistent with the corresponding pH borders, with each metal ion showing sharp chromatographic peaks and completely separated from each other. On the other hand, the peak widths of Eu(III) and Y(III) are larger than those of the other rare earth metal ions because the pH-peak-focusing does not work for the elution of these rare earths. However, compared to the elution peaks obtained by isocratic elution, the peaks for Eu(III) and Y(III) shown in Fig. 7.8 are narrower due to the on-column concentration effect. The use of longer columns and/or a more precise pH gradient program would allow a complete mutual separation of Eu(III) and Y(III).

As shown in Fig. 5.12, the multistep pH-peak-focusing CCC requires 180 min for the separation of La(III), Ce(III), Nd(III), and Yb(III) using a column with a volume of 16.5 mL [4]. In contrast, seven rare earths, including La(III), Ce(III), Nd(III), and Yb(III), can be separated within 100 min by multistep pH-peak-focusing LC with the hydrophilic polymer column, despite a smaller column size (11.8 mL). This indicates that multistep pH-peak-focusing ATPS-LC produces much narrower elution peaks than CCC and can separate metal ions on a preparative scale with shorter columns. Toyopearl HW-40S used in this study is not a high-resolution packing

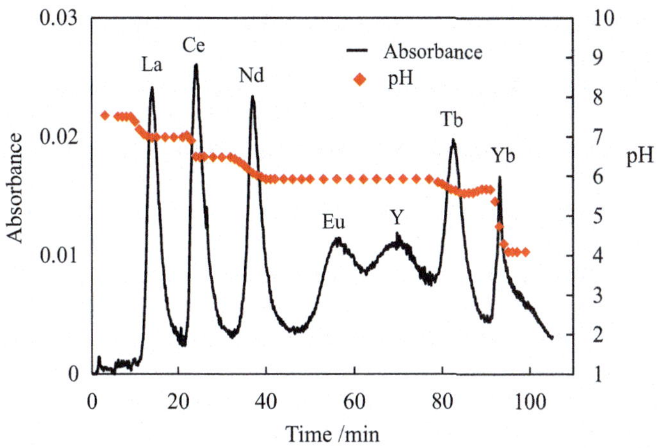

Fig. 7.8 Separation of La(III), Ce(III), Nd(III), Eu(III), Y(III), Tb(III), and Yb(III) by multistep pH-peak-focusing LC using stepwise elution of the mobile phases of which the pH values were adjusted with HEPES (pH 7.5), BES (pH 7.0), MES (pH 6.5, 6.0, and 5.5), and acetate (pH 4.0) buffers. Temperature: 50 °C. See Fig. 7.5 for other chromatographic conditions (reproduced from Ref. [1] by permission of Elsevier)

material for HPLC; the particle size and its distribution are relatively large (20–40 μm). Nevertheless, multistep pH-peak-focusing technique provides the column with high separation performance.

7.3 Partition/Ion-Exclusion Chromatographic Ion Stacking

7.3.1 Mechanism of Ion Stacking in the Process of Partition Chromatography

The separation and concentration of ionic solutes in an aqueous solution is usually achieved by the use of substances with a fixed charge. Ion-exchange resins are probably the most well known of the materials conventionally used to purify and/or concentrate various ions. However, a fixed charge is not a requirement for ion separation materials. A number of nonionic materials are known to be capable of the separation of ionic solutes when used as column packings in LC. Hydrophilic polymer gels, such as cross-linked dextran (Sephadex G gel) and polyacrylamide gel (Bio-gel P gel), are LC packing materials that have long been studied as separation materials for inorganic ions [19–27]. Even hydrophobic materials such as cross-linked polystyrene resin [28–30] and alkyl bonded silica [2, 31–33] have also been reported to exhibit selective ion separation.

In addition to the separation of ions, an interesting finding of ion enrichment has also been reported [2, 3, 28–30]. The increase in concentrations of salts in the aqueous solution containing two different salts was observed when the solution was introduced to a column packed with cross-linked polystyrene beads and eluted with pure water. Figure 7.9 shows a chromatogram obtained by passing a mixture of 4 N $CaCl_2$ and 4 N HCl through a column packed with a cross-linked polystyrene and then replacing the salt solution in the column with pure water [28]. Significant divergence of the elution zones of $CaCl_2$ and HCl and the concentration of $CaCl_2$ and HCl at the front and back edges of the zone can be seen in Fig. 7.9. When this phenomenon was first discovered, it was thought to be caused by the size-exclusion effect from the result that the ions with larger hydrated radii exhibit smaller retention volumes on the column. However, the increase in concentration of ions during elution of the salt solution through the column cannot be ascribed to the size-exclusion mechanism, because the migration velocity of a solute is determined only by its size in size-exclusion chromatography and is constant throughout its migration through the column [2]. The chromatographic behavior of $CaCl_2$ and HCl shown in Fig. 7.9 clearly indicates that the velocity of the salt changes after separation from the mixed salt zone.

As shown in Chap. 6, the separation of solute compounds in an aqueous solution with cross-linked hydrophilic polymer gels is caused by the partition of the solutes into the hydrated polymer gel phase, similar to the polymer-based ATPS. Even on the surface of hydrophobic materials, there exists water that has physicochemical properties and structures different from those of the bulk water and functions as a

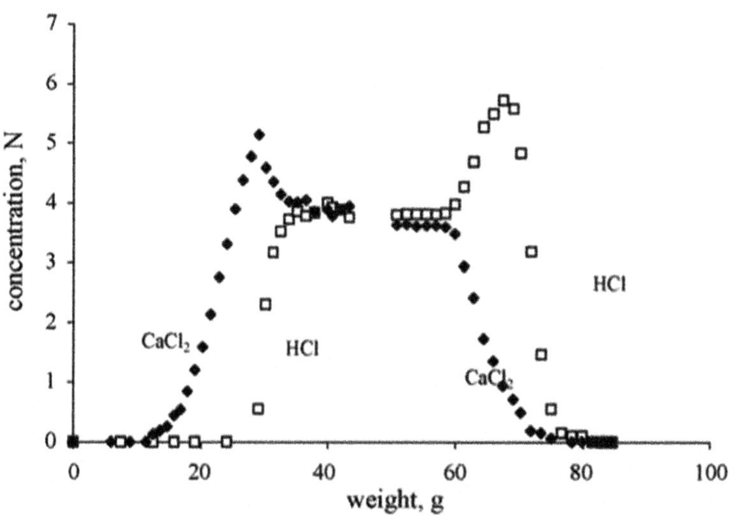

Fig. 7.9 Elution curves of $CaCl_2$ and HCl obtained by elution with water on a hyper-cross-linked polystyrene-type polymer column. The effluent was collected in ca. 1.2 mL fractions using a fraction collector and the weight of the effluent was plotted on the horizontal axes (reproduced from Ref. [28] by permission of Elsevier)

separation medium [31, 32, 34–38]. The increase in concentration of the component salts in a mixed salt solution introduced into the column packed with nonionic materials is attributed to the differential distribution of the ions between the bulk water phase (the mobile phase) and the water incorporated in the polymer matrices (stationary phase).

The distribution of an ionic solute in a two-phase system depends on the background counterion and co-ion due to the charge balance in both phases. Therefore, the distribution equilibria of a cation C^{p+} between the mobile and stationary phases containing an electrolyte consisting of Y^{m+} and X^{n-}, for example, are represented as follows [24, 32]:

$$nC^{p+}(m) + pX^{n^-}(m) \rightleftarrows nC^{p+}(s) + pX^{n^-}(s) \tag{7.1}$$

$$mC^{p+}(m) + pY^{m^+}(s) \rightleftarrows mC^{p+}(s) + pY^{m^+}(m) \tag{7.2}$$

These equilibria are identical with those shown in Chap. 3 (Eqs. (3.4) and (3.5)). When the mobile phase solution of a salt consisting of Y^{m+} and X^{n-} is used, the distribution coefficient, D, for C^{p+} ion obtained on a column packed with nonionic material is thus given by [24, 32]

$$\ln D_C^{YX} = \ln \frac{c_C^{YX}(s)}{c_C^{YX}(m)} = \ln D_C^{WZ} + \frac{p}{m+n}(\ln D_X^{WZ} - \ln D_Y^{WZ}) \tag{7.3}$$

where $c_C^{YX}(m)$ and $c_C^{YX}(s)$ are the concentrations of C^{p+} in the mobile phase and the stationary phase water, and D_C^{WZ}, D_X^{WZ}, and D_Y^{WZ} are the distribution coefficients of C^{p+}, X^{n-}, and Y^{m+} in a mobile phase electrolyte (WZ) system arbitrarily chosen, respectively. The distribution coefficient of an anion A^{p-} in the YX mobile phase electrolyte system is similarly represented by

$$\ln D_A^{YX} = \ln D_A^{WZ} + \frac{p}{m+n}(\ln D_Y^{WZ} - \ln D_X^{WZ}) \tag{7.4}$$

Equations (7.3) and (7.4) show that when the counterions of the background electrolyte in the mobile phase solution are strongly retained, while co-ions are weakly retained, the sample ions will be retained more and the migration rate will be reduced.

The author has also theoretically explained how the distribution coefficients of ionic solutes in partition chromatography vary with the composition of the background electrolyte [39]. When the mobile phase contains univalent anions, X_1^-, ..., X_N^-, and a common cation, Y^+, the distribution coefficients of C^{p+} and A^{p-} are given by the following equations:

$$D_C^{Y(X_1+\cdots+X_N)} = \left(\frac{\sum (D_C^{YX_i})^{2/p}[X_i^-]_m}{\sum [X_i^-]_m} \right)^{p/2} \tag{7.5}$$

$$D_{\mathrm{A}}^{\mathrm{Y}(\mathrm{X}_1+\cdots+\mathrm{X}_N)} = \left(\frac{\sum [\mathrm{X}_i^-]_{\mathrm{m}}}{\sum \frac{[\mathrm{X}_i^-]_{\mathrm{m}}}{(D_{\mathrm{A}}^{\mathrm{YX}_i})^{2/p}}} \right)^{p/2} \tag{7.6}$$

where $[X_i^-]_m$ is the concentration of X_i^- in the mobile phase. Equations (7.5) and (7.6) represent the dependence of the distribution coefficient of a sample ion on the composition of the counterions and on that of the co-ions, respectively. Equation (7.6) gives the following equation for the distribution coefficient of the ion A^- in the mixed electrolyte zone of YA and YB.

$$D_{\mathrm{A}}^{\mathrm{Y}(\mathrm{A}+\mathrm{B})} = \left(\frac{[\mathrm{A}^-]_{\mathrm{m}} + [\mathrm{B}^-]_{\mathrm{m}}}{\frac{[\mathrm{A}^-]_{\mathrm{m}}}{(D_{\mathrm{A}}^{\mathrm{YA}})^2} + \frac{[\mathrm{B}^-]_{\mathrm{m}}}{(D_{\mathrm{A}}^{\mathrm{YB}})^2}} \right)^{1/2} \tag{7.7}$$

If A^- is more weakly retained on the column than B^-, A^- will migrate faster than B^-, resulting in the formation of the YA and YB electrolyte zones at the front and back of the mixed electrolyte zone, respectively, during elution through the column. As soon as the ions A^- and B^- leave the mixed electrolyte (YA + YB) zone, they begin to migrate at the velocities determined by $D_A{}^{YA}$ and $D_B{}^{YB}$, respectively. Since $D_A{}^{YA}$ is larger and $D_B{}^{YB}$ is smaller than $D_A{}^{Y(A+B)}$, YA moves slower in the YA zone than in the mixed electrolyte zone, while YB moves faster in YB zone than in the mixed electrolyte zone. However, the YA and YB zones cannot merge back into the YA + YB zone because $D_A{}^{YA}$ is smaller than $D_B{}^{(YA+YB)}$ and $D_B{}^{YB}$ larger than $D_A{}^{(YA+YB)}$. Electrolytes YA and YB are thus stacked at the leading and trailing edges of the mixed electrolyte zone until the separation of YA and YB is complete.

7.3.2 *Stacking Behavior of Inorganic Anions in Migration Through an Octadecyl-Bonded Silica Column*

Figure 7.10 shows the elution curves obtained by monitoring the absorbance of the column effluent at 235 nm when a mixed solution of equimolar KIO_3 and $KClO_4$ was introduced into an octadecyl-bonded silica column (Capcell Pak C18 UG80) filled with water (a), and followed by elution with pure water (b). Perchlorate and potassium ions show no absorption at this wavelength, so these elution curves represent variations in the iodate ion concentration in the column effluent. As can be seen in Fig. 7.10a, the concentrations of IO_3^- and K^+ associated with IO_3^- are enhanced at the leading edge of the KIO_3 elution zone, while a regular displacement KIO_3 elution curve with water was observed at the trailing edge of the zone, as shown in Fig. 7.10b.

Figures 7.11a and b show the variations in thiocyanate ion concentration in the column effluent when a mixed solution of KCl and KSCN was passed into the

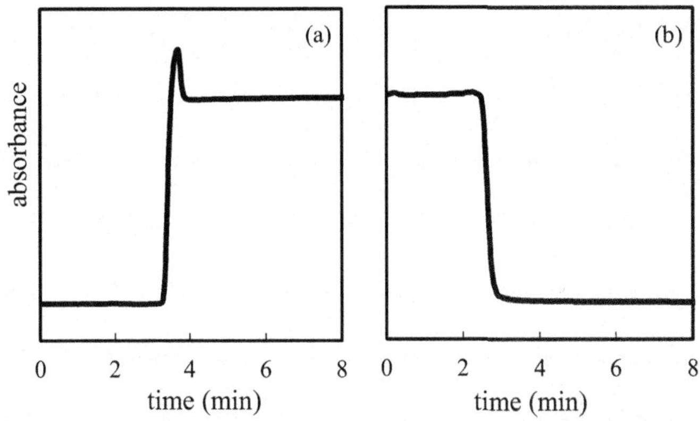

Fig. 7.10 Elution curves of IO_3^- obtained for the introduction of a solution containing KIO_3 and $KClO_4$ to a C_{18} silica column equilibrated with water (**a**) and for the replacement of the solution with pure water (**b**). Column: Capcell Pak C18 UG80 (5 μm; 150 × 4.6 mm; Shiseido, Tokyo, Japan). Concentration of the electrolyte in solution: 1 mmol L^{-1}. Flow rate: 0.5 mL min^{-1}. Detection: UV absorption at 235 nm (reproduced from Ref. [2] by permission of Wiley)

Capcell Pak C18 UG80 column and then eluted with pure water, respectively. These elution curves were obtained by monitoring the absorbance at 235 nm, at which only thiocyanate ion gives rise to absorption. Figure 7.11b shows that the KSCN concentration increases at the trailing edge of its zone, contrary to KIO_3 in a mixture with $KClO_4$.

These ion enrichment phenomena observed on the column packed with nonionic hydrophobic materials cannot be interpreted by the size-exclusion mechanism, but

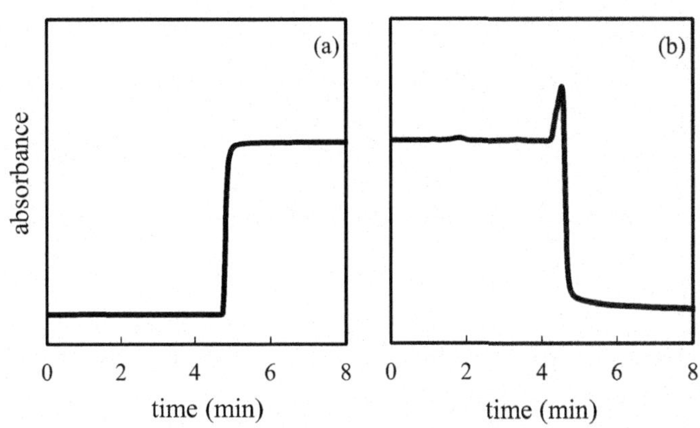

Fig. 7.11 Elution curves of SCN^- obtained for the introduction of a solution containing KCl and KSCN to a C_{18} silica column equilibrated with water (**a**) and for the replacement of the solution with pure water (**b**). For other details, see Fig. 7.10 (reproduced from Ref. [2] by permission of Wiley)

the ion partition model has been successful in explaining them [2, 24, 32, 39]. It has been reported that the interfacial water, which has different structure from that of the bulk water, is formed on the surface of nonionic hydrophobic materials such as polystyrene polymer particles and octadecyl-bonded silica (C_{18} silica), and the separation of ions in aqueous media with C_{18} silica columns can be interpreted as differential ion partitioning between the bulk aqueous phase and the interfacial water [31, 32].

As described in Chap. 6, the volume of the bulk water in a liquid chromatography column, V_{BW}, can be determined using Eq. (6.1) [31, 32, 40]. As the total water volume in the column, V_t, is obtained by measuring the retention volume of deuterium oxide (D_2O), the volume of the interfacial water, which functions as the stationary phase, V_x, is given by $V_t - V_{BW}$. The distribution coefficients of the inorganic anions can thus be calculated as follows:

$$D_A^{YX} = \frac{V_A^{YX} - V_{BW}}{V_x} = \frac{V_A^{YX} - V_{BW}}{V_t - V_{BW}} \tag{7.8}$$

The distribution coefficients for some inorganic anions determined on the Capcell Pak C18 UG80 column with aqueous potassium salt solutions used as the mobile phases are listed in Table 7.1. As chloride and perchlorate ions cannot be detected by a UV–VIS detector, the D values for these ions were calculated using the following equations, which were derived from the rule that the ratio of the distribution coefficients of two ions of the same charge is constant regardless of the type of mobile phase electrolyte [24, 32, 40].

$$D_{Cl}^{KCl} = \frac{D_{Cl}^{KSCN} D_{IO_3}^{KCl}}{D_{IO_3}^{KSCN}} \tag{7.9}$$

$$D_{ClO_4}^{KCl} = \frac{D_{ClO_4}^{KIO_3} D_{SCN}^{KCl}}{D_{SCN}^{KIO_3}} \tag{7.10}$$

Here, the values of D_{Cl}^{KSCN} and $D_{ClO_4}^{KIO_3}$ were obtained by measuring the retention volumes of chloride and perchlorate ions using indirect UV absorption detection [41, 42]. The $D_{IO_3}^{KIO_3}$, $D_{ClO_4}^{KClO_4}$, and D_{SCN}^{KSCN} values were also obtained similarly to the determination of D_{Cl}^{KCl}.

Some of the anions exhibit the distribution coefficients less than 1.0 and appear to be excluded from the pores of the C_{18} silica beads. However, the distribution coefficients for the inorganic ions on the C_{18} silica column are not determined by the sizes of the hydrated ions, but by their relative solubility in the interfacial water and in the bulk mobile phase water [32]. The ions showing small distribution coefficients are strongly hydrated and have weak affinity to the interfacial water. The effect of the background electrolytes in the mobile phase solution on the distribution of an ion is represented by Eqs. (7.3) and (7.4), indicating that the retention of an ionic solute in

Table 7.1 Distribution coefficients of inorganic anions on a Capcell Pak C18 UG80 with 0.01 mol L^{-1} potassium salt solutions (reproduced from Ref. [2] by permission of Wiley)

Anion (A)	$D_A{}^{KCl}$	$D_A{}^{KIO3}$	$D_A{}^{KClO4}$	$D_A{}^{KSCN}$
$IO_3{}^-$	0.60	(0.63)[a]	0.37	
Cl^-	(0.66)[a]			0.37
Br^-	0.74			
$NO_3{}^-$	0.90			
I^-	1.07			
$ClO_4{}^-$	(1.57)[a]	1.72	(1.01)[a]	
SCN^-	1.99			(1.11)[a]

[a]Calculated values

aqueous liquid chromatography with a column packed with a nonionic material can rationally be interpreted using the ion partition model [24, 32].

The distribution coefficient of an anion during migration in a mixed electrolyte zone can be calculated by inserting the D values shown in Table 7.1 into Eq. (7.7). The distribution coefficients calculated for $IO_3{}^-$ and $ClO_4{}^-$ in the solution zone of an equimolar mixture of KIO_3 + $KClO_4$ and those for Cl^- and $ClO_4{}^-$ in the KCl + KSCN zone are listed in Table 7.2. $IO_3{}^-$ and $ClO_4{}^-$ move in the KIO_3 + $KClO_4$ mixed electrolyte zone at the velocities determined by $D_{IO_3}^{K(IO_3+ClO_4)}$ (= 0.45) and $D_{ClO_4}^{K(IO_3+ClO_4)}$(= 1.23), respectively. Thus, KIO_3 migrates faster than $KClO_4$ and separates from the leading edge of the mixed electrolyte zone, while $KClO_4$ separates from the trailing edge of the zone during elution through the column. After the separation, they migrate as pure KIO_3 and $KClO_4$ solution zones at the velocities in accordance with $D_{IO_3}^{KIO_3}$(= 0.63) and $D_{ClO_4}^{KClO_4}$(= 1.01), respectively. Consequently, the KIO_3 and $KClO_4$ zones continue to decrease in width until completely separated from each other, thus leading to an increase in concentration.

Similarly, the introduction of the equimolar mixture of KCl and KSCN into the column followed by the elution with pure water results in an increase in the concentration of KCl at the front of the mixed electrolyte zone and of KSCN at the rear edge of the zone. Although the elution profiles of KCl and $KClO_4$ are not experimentally obtained by UV detection, $KClO_4$ is concentrated at the trailing edge of the KIO_3

Table 7.2 Calculated distribution coefficients of the component anions in the solution zones of equimolar mixture of KIO_3 + $KClO_4$ and KCl + KSCN on a Capcell Pak C18 UG80 column (reproduced from Ref. [2] by permission of Wiley)

Anion	$D_A{}^{K(IO3+ClO4)}$	$D_A{}^{K(Cl+SCN)}$
$IO_3{}^-$	0.45	
$ClO_4{}^-$	1.23	
Cl^-		0.45
SCN^-		1.37

zone shown in Fig. 7.10b and KCl is accumulated at the leading edge of the KSCN zone shown in Fig. 7.11a.

The degree of the enrichment of an ion A^- at the leading or trailing edge of the sample electrolyte solution zone depends on the difference in the distribution coefficient between the ion A^- and the coexisting co-ion B^-; the larger the difference, the greater the enrichment. The elution curves obtained for the introduction of KIO_3, KNO_3, KI, and KSCN solutions containing equimolar $KClO_4$ to the Capcell Pak C18 UG80 column are shown in Fig. 7.12. The order of the degree of the ion enrichment at the leading edge of the elution zone is $IO_3^- > NO_3^- > I^- > SCN^-$, which corresponds to the order of the difference in D value between the ion and perchlorate ion (Table 7.1). Thiocyanate ion does not show stacking at the leading edge of the elution zone because it has a larger D value than perchlorate ion. The results described above indicate that the ion-stacking behavior observed when a mixed electrolyte solution is introduced into a column packed with a nonionic material can be well explained by the ion partition model.

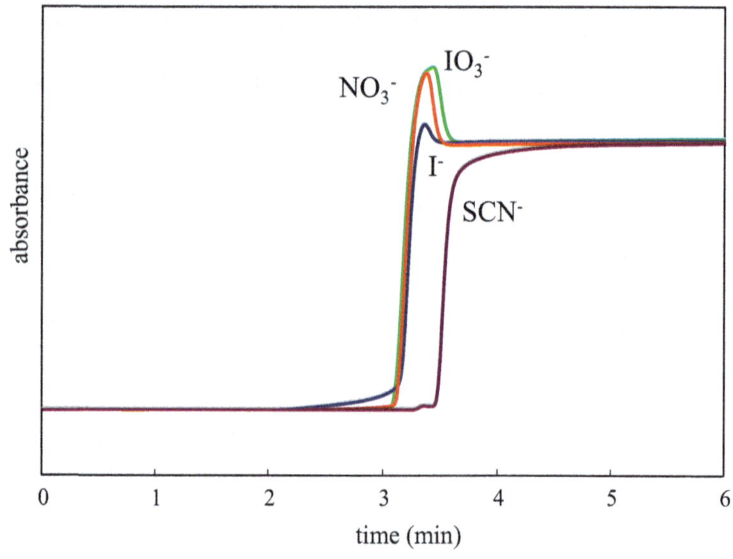

Fig. 7.12 Elution curves obtained for the introduction of KIO_3, KNO_3, KI, and KSCN solutions containing equimolar $KClO_4$ to a C_{18} silica column equilibrated with water. Absorbance values are normalized to the values for the sample solutions introduced. Column: Capcell Pak C18 UG80 (5 μm, 150 × 4.6 mm). Concentration of the electrolyte in solution: 1 mmol L^{-1}. Flow rate: 0.5 mL min^{-1}. Detection: UV absorption at 235 nm (reproduced from Ref. [2] by permission of Wiley)

7.3.3 *Ion Stacking by a Combination of Partition and Ion Exclusion with a Hydrophilic Polymer Gel Column*

The results presented in the previous chapter indicate that trace ionic species in aqueous solutions can be readily concentrated by adding an appropriate electrolyte to the solution, introducing the solution into a column packed with a nonionic material, and then eluting with pure water. However, practical ion enrichment columns require packing materials that provide both high retention capacity and sufficient selectivity for different ions. Many types of hydrophilic polymer gels show good retention selectivity for inorganic ions [5–7, 43]. Hydrophilic polymer gels have a higher retention capacity than hydrophobic materials because they can sorb a larger amount of water, which in turn acts as the stationary phase to retain the ions. A cross-linked hydroxylated methacrylic polymer gel column, TSKgel G2500PW$_{XL}$ (6 μm; 300 mm × 7.8 mm; Tosoh, Tokyo, Japan), was thus adopted for the investigation of the enrichment of inorganic anions in aqueous solutions. The values of V_t, V_{BW}, V_x, and the interstitial or interparticulate void volume, V_{int}, for the TSKgel G2500PW$_{XL}$ column are shown in Table 7.3 in comparison with the values for the Capcell Pak C18 UG80 column. The ratio of the stationary phase water volume to the total column volume, V_x/V_t, is much larger for the TSKgel G2500PW$_{XL}$ column than for the Capcell Pak C18 UG80 column, which means that the former column may have a higher resolution than the latter as described in Chap. 5.

Organic polymer packing materials usually have fixed anionic charges on their matrices, thought of as carboxyl groups, and exhibit ion exclusion for anionic solutes under neutral and alkaline conditions at a low ionic strength [20, 40]. However, because the amount of the fixed charges is relatively small, at a higher ionic strength, the ion-exclusion effect is suppressed and the partition between the bulk water phase (mobile phase) and the polymer gel phase (stationary phase) determines the retention of anions on the column. This means that, if a target anion has the distribution coefficient larger than the co-anion of the background coexisting electrolyte, the elution band of the target anion accumulated at the trailing edge of the electrolyte zone does not expand even after the separation from the background electrolyte zone, because the ion exclusion overcomes the partition and accelerates the migration of the target anion. Consequently, it is expected that combining partition and ion-exclusion effects allows for higher anion enrichment.

Table 7.3 Total water volume in the column, V_t, bulk water volume, V_{BW}, stationary phase water volume, V_x, and the interstitial volume, V_{int} (mL) determined for a TSKgel G2500PW$_{XL}$ column and a Capcell Pak C18 UG80 column (reproduced from Ref. [2] by permission of Wiley)

	V_t	V_{BW}	V_x	V_{int}
TSKgel G2500PW$_{XL}$	10.22 ± 0.04	6.80 ± 0.18	3.42 ± 0.18	5.27 ± 0.05
Capcell Pak C18 UG80	1.47	1.24 ± 0.03	0.23 ± 0.03	

Figure 7.13 shows the variation of the retention volumes of NO_3^-, I^-, and SCN^- on the TSKgel G2500PW$_{XL}$ column with the concentration of NaCl in the mobile phase solution. It can be seen that, at the NaCl concentration of 1 mmol L^{-1} or less, all the anions have retention volumes close to the V_{int} value of the column (= 5.27 mL), but with an increase in NaCl concentration, the retention volumes of the anions gradually increase and reach their respective constant values at NaCl concentration of 0.1 mol L^{-1} or above. The amount of the fixed anionic charges or cation-exchange capacity for the TSKgel G2500PW$_{XL}$ column was determined to be 3.4 μeq mL^{-1}. These experimental results demonstrate that anions are excluded out of water-swollen TSKgel G2500PW$_{XL}$ particles by the ion exclusion at an ionic strength of 1 mmol L^{-1} or below, whereas at 0.1 mol L^{-1} or above, the exclusion effect is suppressed and the anions can freely distribute into the polymer gel phase.

The distribution coefficients for several inorganic anions when using 0.1 mol L^{-1} NaCl and Na_2SO_4 as the mobile phases are given in Table 7.4. The difference in the distribution coefficients among the anions obtained with the TSKgel G2500PW$_{XL}$ column is much larger than that obtained with the Capcell Pak C18 UG80 column, which indicates that a higher enrichment of target ions can be expected when using the TSKgel G2500PW$_{XL}$ column due to both the partition and ion-exclusion effects.

Figure 7.14a shows the variation of the chromatogram of nitrate ion with the injection volume obtained when the sample solution containing 1 μmol L^{-1} $NaNO_3$ and 0.1 mol L^{-1} NaCl is injected and eluted with pure water. A very sharp peak was observed and the peak height increased with increasing sample volume, maintaining the same peak width until the sample injection volume reached 1.9 mL. Without the

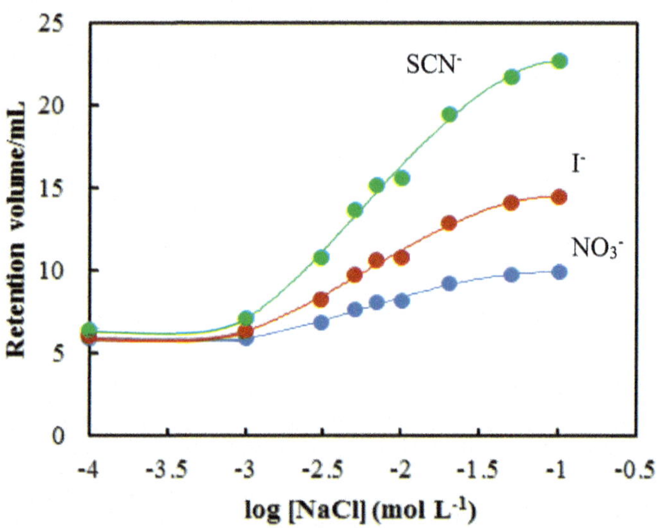

Fig. 7.13 Dependence of the retention volumes of NO_3^-, I^-, and SCN^- on the concentration of the mobile phase electrolyte (NaCl). Column: TSKgel G2500PW$_{XL}$. (reproduced from Ref. [2] by permission of Wiley)

Table 7.4 Distribution coefficients of inorganic anions on a TSKgel G2500PW$_{XL}$ column with 0.1 mol L^{-1} NaCl and Na_2SO_4 solutions (reproduced from Ref. [2] by permission of Wiley)

Anion (A)	D_A^{NaCl}	$D_A^{Na_2SO_4}$
SO_4^{2-}	(0.11)[a]	(0.32)[a]
IO_3^-	0.20	0.42
Cl^-	(0.36)[a]	(0.60)[a]
Br^-	0.71	1.42
NO_3^-	0.93	1.80
I^-	2.27	4.42
SCN^-	4.66	8.99

[a]calculated values

ion-exclusion effect, although some enrichment occurs due to the partition effect, the width of the nitrate zone would increase as the volume of the solution injected increases. This is because the pure $NaNO_3$ zone, separated from the NaCl zone, moves slower than the trailing edge of NaCl zone ($D_{NO_3}^{NaNO_3} > D_{Cl}^{NaCl}$).

However, the peaks obtained by injecting 2.7 mL or 5.0 mL of the solution are accompanied by a broadened leading zone and the enrichment is incomplete. It should be noted that the flat leading zone starts at the retention time for the nitrate ion peak obtained by injection of 1.9 mL of the solution. This indicates that nitrate ions accumulate at the trailing edge of the NaCl zone and are eluted from the column in a narrow zone by the ion-exclusion effect. However, if the volume of sample solution exceeds 1.9 mL, nitrate ions will be eluted from the column before they are completely separated from the NaCl zone; nitrate ions remaining in the NaCl zone show a flat zone in the chromatogram, the height of which corresponds to the concentration in the sample solution (1 μmol L^{-1}).

The incomplete stacking of nitrate ions when more than 1.9 mL of solution is injected is due to the small difference in the distribution coefficients of NO_3^- and Cl^- in the NaCl zone. The retention volume of the leading edge of the eluting zone of NO_3^-, before NO_3^- is completely separated from the zone of NaCl, is equal to that of NO_3^- obtained using NaCl solution as the mobile phase, and is given by

$$V_{NO_3}^{NaCl} = V_{BW} + D_{NO_3}^{NaCl} V_X \quad (7.11)$$

On the other hand, the retention volume of the trailing edge of the NaCl zone, $V_{tr(NaCl)}$, is represented by the following equation since the concentration of NO_3^- is negligibly smaller than Cl^-.

$$V_{tr(NaCl)} = V_{BW} + D_{Cl}^{NaCl} V_X + V_{smp} = V_{Cl}^{NaCl} + V_{smp} \quad (7.12)$$

where V_{smp} is the injection volume of the sample solution. Equations (7.11) and (7.12) indicate that complete stacking of NO_3^- can be accomplished when $V_{smp} \le V_{NO_3}^{NaCl} - V_{Cl}^{NaCl}$. If the sample volume is so large that $V_{tr(NaCl)}$ exceeds $V_{NO_3}^{NaCl}$, or $V_{smp} > V_{NO_3}^{NaCl} - V_{Cl}^{NaCl}$, the complete accumulation of NO_3^- in a single sharp zone will

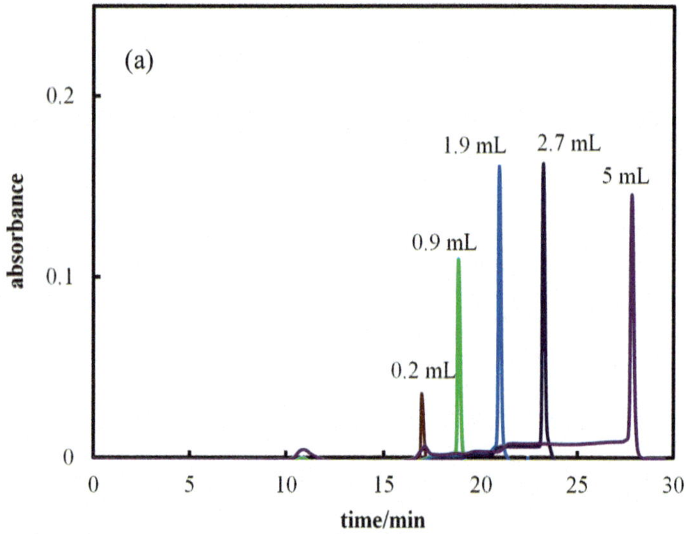

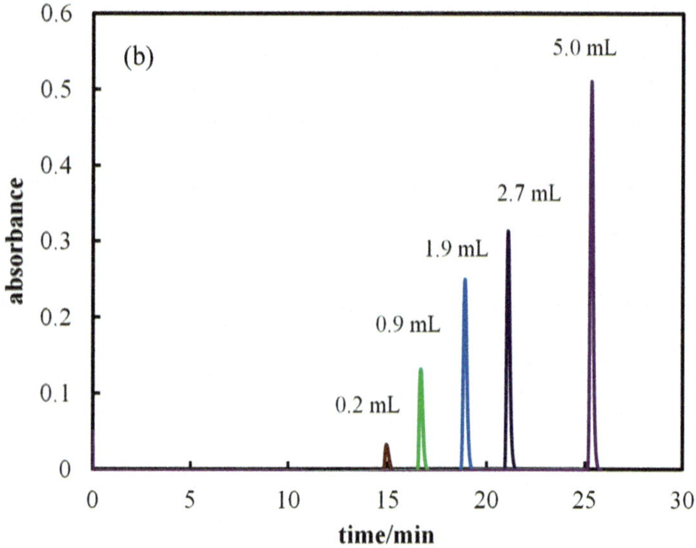

Fig. 7.14 Variation of the chromatogram of NO_3^- obtained by injecting mixed electrolyte solutions of $NaNO_3$ + NaCl (**a**) and $NaNO_3$ + Na_2SO_4 (**b**) with the sample volume injected. Column: TSKgel G2500PW_{XL}. Concentration of electrolyte in the sample solution: $NaNO_3$ = 1 μmol L^{-1}; NaCl and Na_2SO_4 = 0.1 mol L^{-1}. Flow rate: 0.5 mL min^{-1} (reproduced from Ref. [2] by permission of Wiley)

not be able to be performed. The retention volumes for Cl^- and NO_3^- in 0.1 mol L^{-1} NaCl mobile phase system, $V_{NO_3}^{NaCl}$ and V_{Cl}^{NaCl}, are calculated to be 9.98 and 8.03 mL, respectively, from the V_{BW}, V_X, and D^{NaCl} values shown in Tables 7.3 and 7.4. Thus, when NaCl is added as the background electrolyte, the upper limit of the sample solution volume for complete enrichment of nitrate ion is calculated to be 1.95 mL, which is consistent with the results shown in Fig. 7.14a.

The chromatogram of nitrate ion changes as shown in Fig. 7.14b when the added electrolyte is changed to sodium sulfate. Nitrate ion shows a sharp elution peak even when 5.0 mL of the sample solution is injected. This result is due to the fact that $D_{NO_3}^{Na_2SO_4}$ is much larger than $D_{SO_4}^{Na_2SO_4}$ (Table 7.4), indicating that in order to obtain a higher concentration of the target anion, an electrolyte with an anion of which the distribution coefficient is sufficiently smaller than that of the target anion should be added to the sample solution.

To quantitatively evaluate the degree of enrichment obtained by the present method, partition/ion-exclusion chromatographic ion stacking (PIEC ion stacking), the enrichment factor, E, defined by the following equation, was calculated.

$$E = \frac{A_{peak}}{A_{smp}} \tag{7.13}$$

where A_{peak} is the absorbance at the top of the peak of a target ion and A_{smp} is the absorbance shown by the target ion in the sample solution. Figure 7.15 shows the E values plotted against the injection volume of the sample solution for nitrate, iodide, and thiocyanate ions and compares the effects of NaCl and Na_2SO_4 as a background electrolyte added to the sample solution on the E value. All the plots, with the exception of the NaCl-added nitrate ion solution, show straight lines with the same slope (ca. 12 mL^{-1}). This means that the stacked target ion zones do not disperse as they migrate through the column and maintain a constant width regardless of sample injection volume.

The upper limits of the volume of the sample solution for complete enrichment of nitrate, iodide, and thiocyanate ions by the addition of Na_2SO_4 are calculated to be 5.0, 14.0, and 29.7 mL, respectively, from the retention volumes of the anions shown in Table 7.4. Therefore, it is predicted that the enrichment factors for these anions will be increased up to 60, 170, and 350 for nitrate, iodide, and thiocyanate ions, respectively. The enrichment of a trace amount of nitrate ion in a high concentration Na_2SO_4 solution by PIEC ion stacking is schematically depicted in Fig. 7.16. PIEC ion stacking is a simple method that can easily achieve over a 100-fold enrichment of ionic solutes, provided that the ionic strength of the accumulated target ion zone remains within the range where the ion-exclusion effect is predominant (1 mmol L^{-1} or less for the TSKgel G2500PW$_{XL}$ column).

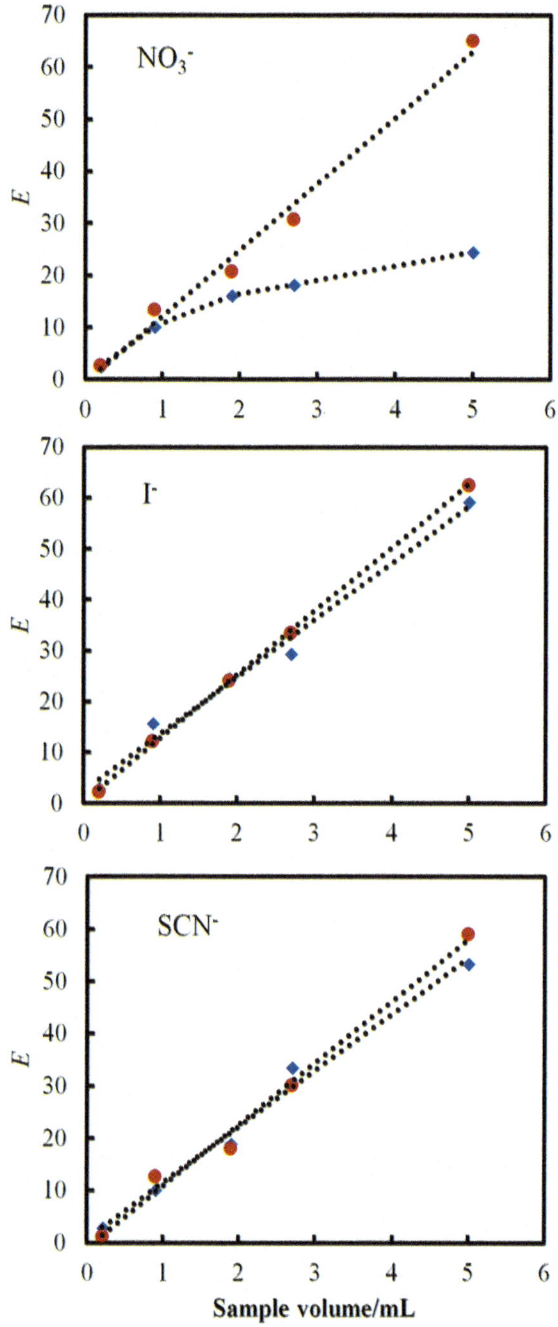

Fig. 7.15 Relationships between the enrichment factor and the sample volume injected for NO_3^-, I^-, and SCN^-. Column: TSKgel G2500PW$_{XL}$. Mobile phase: pure water. Background electrolyte: ● = Na_2SO_4; ◆ = NaCl. Concentration of the target ion in the sample solution = 1 μmol L^{-1}. Concentration of the added background electrolyte in the sample solution = 0.1 mol L^{-1} (reproduced from Ref. [2] by permission of Wiley)

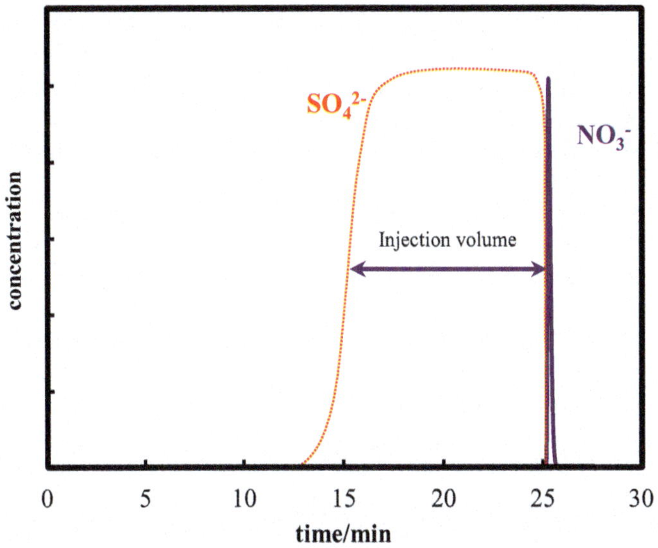

Fig. 7.16 Schematic representation of partition/ion-exclusion chromatographic ion stacking for nitrate ion in a solution of sodium sulfate (reproduced from Ref. [3] by permission of Springer Nature)

7.3.4 Analysis of Trace Amounts of Anions in Salt Samples by Online PIEC Ion Stacking-Ion Chromatography

The determination of trace inorganic anions in samples is a common analytical problem especially when the samples contain large excesses of matrix salts [44–46]. In order to accurately and precisely determine trace anions in salt matrices, the analyte anions must be selectively separated from the matrix and concentrated prior to analysis. Among various analytical methods that have been reported for the determination of anionic solutes [47–52], ion chromatography (IC) is one of the most accepted techniques for determination of a wide range of inorganic anions in water samples, because it has the advantages of high-speed analysis and high resolution [44–46, 53–56]. In the determination of trace anions by IC, a preconcentrator column is often used to concentrate the analyte anions from large volume of sample solution to achieve sufficient sensitivities [57–61]. However, to obtain accurate and precise results, the loading of the analyte ions on the column and elution of them from the column must be highly reproducible. In addition, the eluent used to elute the analyte anions from the column should not interfere with the separation and detection of the analyte ions in the IC system.

PIEC ion stacking is an ideal method for preconcentrating anionic analytes for IC analysis because the effective enrichment can be achieved only by adding an appropriate electrolyte to the sample solution, injecting the solution into a hydrophilic polymer column, and eluting it with pure water. The hyphenation of PIEC ion stacking and IC, online PIEC ion stacking-IC, thus provides a powerful method for the determination of trace amounts of anions in a sample. In particular, online PIEC ion stacking-IC is a useful method for the analysis of trace impurity anions in salt samples. In this section, it will be demonstrated that online PIEC ion stacking-IC can be successfully applied to the determination of trace amounts of anions in reagent-grade salts.

Figure 7.17 shows the online PIEC ion stack-IC system assembled in our laboratory. The preconcentration of analyte anions is performed using the first dimension HPLC system consisting of an HPLC pump, a sample injector equipped with a 1.9 or 5.0 mL loop, the anion concentrator column, TSKgel G2500PW$_{XL}$ (7.8 × 300 mm), and a UV detector. The second dimension IC system used for determination of trace anions is a Dionex Model ICS-1600 (California, USA) consisting of a pump, an AS 22 guard column (4 × 50 mm), an AS 22 separator column (4 × 250 mm), and a UV detector. The first and second dimension systems are connected by a 10-port motorized valve. The mobile phase used for the separation of anionic analytes in the IC system is aqueous solution containing 4.5 mmol L^{-1} Na_2CO_3 and 1.4 mmol L^{-1} $NaHCO_3$. Analyte anions eluted from the concentrator column were collected in a 298 μL loop attached to the 10-port valve and then transported to the second IC system. The valve switching time was optimized so that the column eluate, including the entire concentration zone, was collected in the receiver loop.

The maximum injection volume of the sample solution, V_{smp}^{Max}, for a complete stacking of an analyte ion using the PIEC stacking method is derived from Eqs. (7.11) and (7.12) and is given as the difference in the retention volumes of the analyte anion

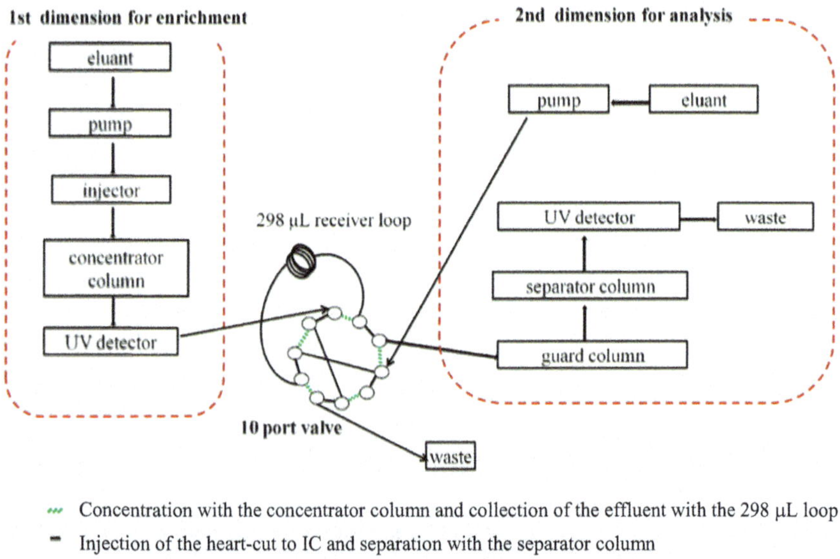

Fig. 7.17 Instrumental set up of online PIEC ion stacking-IC (reproduced from Ref. [3] by permission of Springer Nature)

A^- and the anion X^- of the background electrolyte YX, by the following equation:

$$V_{smp}^{Max} = V_A^{YX} - V_X^{YX} = (D_A^{YX} - D_X^{YX})V_x \tag{7.14}$$

where $V_A^{YX} > V_X^{YX}$. The width of the elution zone of the stacked analyte anion is constant regardless of the injection volume of the sample solution. Therefore, the smaller the distribution coefficient of the component anion X^- of the background electrolyte, the higher the enrichment of the analyte anions obtained by PIEC stacking. As sulfate ion has the smallest distribution coefficient among the common inorganic anions (Table 7.4), Na_2SO_4 was chosen as the electrolyte to be added to the sample solution. Substituting the values of $D_{SO_4}^{Na_2SO_4}$ (0.32) and V_x of the TSKgel G2500PW$_{XL}$ column (3.42 mL) into Eq. (7.14), it is expected that for anions with the distribution coefficients greater than 1.78 in the Na_2SO_4 mobile phase system, the complete anion stacking is possible even when loading up to 5.00 mL of sample solution.

The ion-exclusion effect on the elution of anions exerted by a TSKgel G2500PW$_{XL}$ column is suppressed at an ionic strength of 0.1 mol L^{-1} or above (Fig. 7.13) and the anions can be retained in the stationary gel phase. Therefore, the concentration of Na_2SO_4 was set to 0.1 mol L^{-1} (ionic strength = 0.3 mol L^{-1}).

First, the reproducibility of the online PIEC ion stacking-IC method was verified by determining nitrate ion in tap water. Figure 7.18 shows the chromatogram obtained by elution with pure water for an injection of 5.00 mL of tap water diluted 50-fold with 0.100 mol L^{-1} Na_2SO_4 into the concentrator column. To minimize the contamination of nitrate ion from the added salt, Na_2SO_4 of Suprapur (SP) grade (Merck, Darmstadt, Germany) was used. The sharp single peak shown in Fig. 7.18 is attributed to the anions showing absorption at 220 nm in the tap water sample, which have accumulated at the trailing edge of the Na_2SO_4 elution zone.

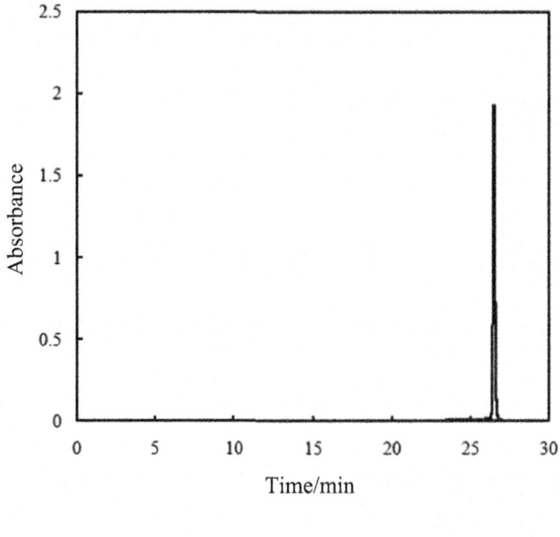

Fig. 7.18 Chromatogram obtained for an injection of 50-fold diluted tap water with 0.100 mol L^{-1} Na_2SO_4 into the concentrator column. Injection volume: 5.00 mL, mobile phase: pure water, flow rate: 0.5 mL min^{-1}, detection wavelength: 220 nm (reproduced from Ref. [3] by permission of Springer Nature)

Figure 7.19a shows a chromatogram obtained by the IC of the online PIEC ion stacking-IC system. The peak for nitrate ion was detected in front of a negative peak, which is due to the elution of sulfate ion present together with nitrate ion in the fraction heart-cut from the concentrator effluent. The chromatogram obtained by a direct injection of 298 μL of tap water diluted 50-fold with pure water to the IC system is shown in Fig. 7.19b. As can be seen, the area of nitrate ion peak obtained by online PIEC ion stacking-IC is significantly larger than that obtained by IC due to the enrichment by PIEC ion stacking.

The results of the spike recovery test of nitrate ion performed on injections of 5.00 mL tap water sample diluted 50-fold with 0.100 mol L^{-1} Na_2SO_4 are shown in Table 7.5. The recovery is satisfactory and the low standard deviation of the measurement values is an indication of the high precision of the present method. The detection limit, defined as three-fold of the standard deviation of the blank signal ($n = 3$), was 0.5 μg L^{-1}, which indicates that the online PIEC ion stacking-IC method enables one to determine ppb level of trace anions in sample solutions.

It is expected that online PIEC ion stacking-IC is a useful method for determination of trace impurity anions in salt samples because trace anionic contents can be measured by injecting the solutions prepared by dissolving the salts in water directly into the online PIEC ion stacking-IC system. It was thus attempted to

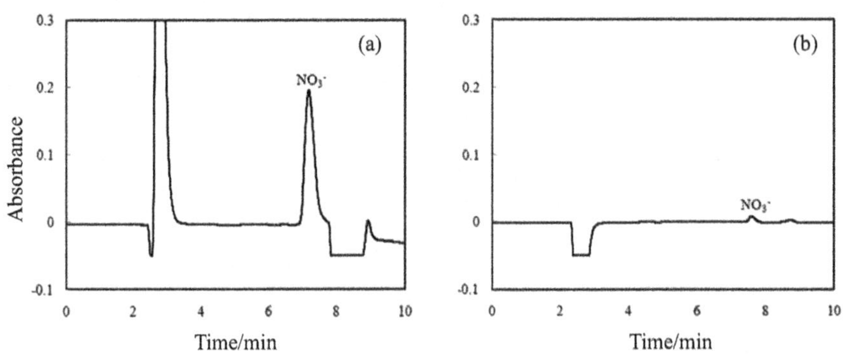

Fig. 7.19 Chromatograms obtained by injection of 50-fold diluted tap water with 0.100 mol L^{-1} Na_2SO_4 into the online PIEC ion stacking-IC (**a**) and direct injection of 50-fold diluted tap water with water into IC (**b**) Mobile phase, 4.5 mmol L^{-1} Na_2CO_3 + 1.4 mmol L^{-1} $NaHCO_3$; flow rate, 1.2 mL min^{-1}; detection wavelength, 220 nm (reproduced from Ref. [3] by permission of Springer Nature)

Table 7.5 Results of the recovery test for nitrate determination ($n = 3$) in a 50-fold diluted tap water (reproduced from Ref. [3] by permission of Springer Nature)

Sample	Ion	Added (mg L^{-1})	Found (mg L^{-1})	Recovery (%)
Tap water	NO_3^-	0	0.122 ± 0.001	–
		0.050	0.173 ± 0.002	105 ± 2
		0.100	0.219 ± 0.001	97 ± 2

determine trace impurity anions in Japanese Industrial Standards (JIS) guaranteed reagent (GR)-grade and extra pure (EP)-grade salts of Na_2SO_4 by online PIEC ion stacking-IC.

Figure 7.20 shows the chromatograms obtained for injections of 0.100 mol L^{-1} aqueous solutions of the two different grade salts of Na_2SO_4. The peaks for nitrate and nitrite ions were detected for both GR- and EP-grade salts. The results of spike recovery tests are summarized in Table 7.6. The V_{smp}^{Max} value for nitrite ion calculated from its distribution coefficient ($D_{NO_2}^{Na_2SO_4} = 1.62$) according to Eq. (7.14) is 4.45 mL. Therefore, the complete stacking of nitrite ion is not achieved with the concentrator column used. However, the mass fraction of nitrite ion collected in the receiver loop relative to the total amount of nitrite is calculated as 95% (= (4.45 + 0.30)/5.00). As shown in Table 7.6, the recovery values were 97–105% for the measurements of nitrate ion, while the values for nitrite determination are around 95% as predicted.

The use of pure water for enrichment of analyte anions by PIEC ion stacking results in extremely low background levels. The detection limits for nitrite and nitrate

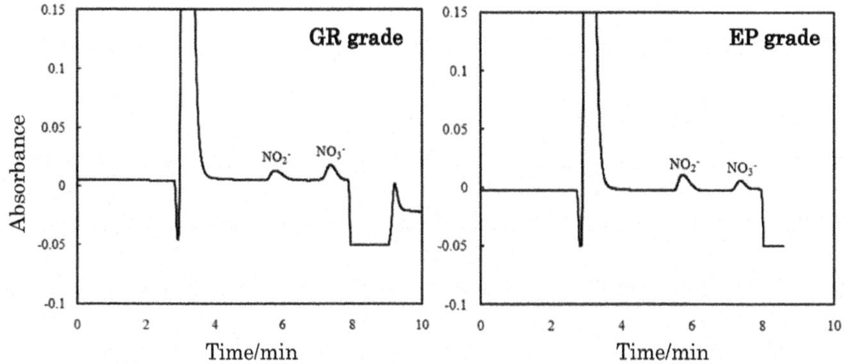

Fig. 7.20 Chromatograms obtained for the injections of 0.100 mol L^{-1} GR- and EP-grade Na_2SO_4 solutions by online PIEC ion stacking-IC. For other details see Fig. 7.19 (reproduced from Ref. [3] by permission of Springer Nature)

Table 7.6 Results of the recovery tests for nitrate and nitrite determination ($n = 3$) in 0.100 mol L^{-1} Na_2SO_4 solution (reproduced from Ref. [3] by permission of Springer Nature)

Ion	Added (μg/L)	GR grade		EP grade	
		Found (μg/L)	Recovery (%)	Found (μg/L)	Recovery (%)
NO_3^-	0	10.3 ± 0.1	–	6.7 ± 0.2	–
	10	20.5 ± 0.1	102 ± 1	17.2 ± 0.5	105 ± 4
	20	31.4 ± 0.3	105 ± 1	26.1 ± 0.4	97 ± 2
NO_2^-	0	6.9 ± 0.1	–	13.0 ± 0.5	–
	5	11.6 ± 0.4	94 ± 8	17.8 ± 0.6	96 ± 12
	10	16.5 ± 0.3	96 ± 3	26.1 ± 0.4	97 ± 3

Table 7.7 Quantification results of GR- and EP-grade sodium sulfate for impurity $NaNO_2$ and $NaNO_3$ (reproduced from Ref. [3] by permission of Springer Nature)

Salt	GR grade/mg kg^{-1}	EP grade/mg kg^{-1}
$NaNO_3$	1.01 ± 0.01	0.66 ± 0.01
$NaNO_2$	0.73 ± 0.01	1.39 ± 0.05

ions in the analysis of reagent-grade Na_2SO_4 salt solutions were 0.02 μg L^{-1} and 0.04 μg L^{-1}, respectively. Table 7.7 shows the contents of impurity nitrite and nitrate in the GR- and EP-grade salts of Na_2SO_4 calculated as the mass percentages of $NaNO_2$ and $NaNO_3$ in the sodium salt samples assuming that the recoveries of both of the anions are 100%. These impurities were found to be much smaller than the limits specified by the manufacturers. These results demonstrate that trace amounts of impurity anions present even at sub-ppm levels in salt samples can be determined by online PIEC ion stacking-IC.

7.4 Conclusions

Liquid chromatography using a hydrophilic polymer gel column and an aqueous mobile phase is, in principle, a separation method based on polymer-based ATPS. ATPS-LC provides unique methods for the separation and/or enrichment of ionic solutes because both the mobile phase and the stationary phase are essentially made up of water and can accommodate ions in their intact hydrated form. Multistep pH-peak-focusing LC and PIEC ion stacking are ATPS-LC methods developed for the separation of metal ions and enrichment of anionic solutes utilizing the characteristics of ATPS, respectively.

It was demonstrated that multistep pH-peak-focusing LC using a hydroxylated methacrylic polymer gel column and an aqueous Na_2SO_4 mobile phase solution containing acetylacetone and 1,10-phenanthroline as the extractants is useful for simultaneously separating and concentrating rare earth metal ions. The column is first equilibrated with the mobile phase of pH at which all sample metal ions are retained on the column by synergistic extraction due to complexation with the two extractants. After injection of the sample solution, the mobile phases, the pH values of which have been adjusted with appropriate buffer reagents, are pumped into the column by stepwise gradient elution in order of decreasing pH. At a pH border formed between different pH zones in the LC column, each metal ion is enriched. The mutual separation of La(III), Ce(III), Nd(III), Eu(III), Y(III), Tb(III), and Yb(III) was achieved by multistep pH-peak-focusing LC. Unlike CCC with polymer-based ATPS, ATPS-LC allows fast and highly efficient separation and prevents polymer loss in practical applications because the mobile phase is an inorganic salt solution and does not contain a polymer. Multistep pH-peak-focusing LC with a hydrophilic polymer gel column therefore has great potential as a useful, environmentally friendly method for the simultaneous enrichment and separation of metal ions.

When an aqueous mixed electrolyte solution is injected into a column packed with a nonionic material and eluted with pure water, the ions exhibiting weaker retention than the other ions in the solution are stacked at the leading edge of the mixed electrolyte zone, while those showing stronger retention are accumulated at the trailing edge. The PIEC ion stacking method was conceived and developed as a result of a theoretical investigation of this phenomenon. Based on the combination of partition chromatographic ion stacking with ion exclusion by fixed anionic charges on a hydrophilic polymer gel, this method can achieve highly efficient enrichment of trace amounts of anions in samples. A new method for the analysis of trace anions, online PIEC ion stacking-IC, which combines PIEC ion stacking and ion chromatography, was proposed and shown to be capable of determining trace nitrite and nitrate at sub-ppm levels in reagent-grade sodium sulfate salts.

The development of multistep pH-peak-focusing LC and PIEC ion stacking has only just begun and is far from complete. The use of columns with higher column efficiency and selectivity suitable for each method will not only increase the efficiency of separation and enrichment, but also expand the range of applications.

References

1. Shibukawa M, Onoyama Y, Handa-Tasaki Y, Saito S (2024) Multistep pH-peak-focusing liquid chromatography with a hydrophilic polymer gel column for separation of rare earth elements. J Chromatogr A 1721:464829
2. Akter F, Ogiyama Y, Saito S, Shibukawa M (2017) Mechanism of ion stacking in aqueous partition chromatographic processes. J Sep Sci 40:3205–3213
3. Akter F, Saito S, Handa-Tasaki Y, Shibukawa M (2018) Partition/ion-exclusion chromatographic ion stacking for analysis of trace anions in water and salt samples by ion chromatography. Anal Sci 34:369–373
4. Shimizu K, Kuribayashi H, Watanabe H, Shimasaki T, Azuma K, Horie Y, Saitoh K, Saito S, Shibukawa M (2013) Multistep pH-peak-focusing countercurrent chromatography with a polyethylene glycol-Na_2SO_4 aqueous two phase system for separation and enrichment of rare earth elements. Anal Chem 85:978–984
5. Baba T, Sakamoto R, Shibukawa M, Oguma K (2004) Solute retention and the states of water in polyethylene glycol and poly(vinyl alcohol) gels. J Chromatogr A 1040:45–51
6. Shibukawa M, Aoyagi K, Sakamoto R, Oguma K (1999) Liquid chromatography and differential scanning calorimetry studies on the states of water in hydrophilic polymer gel packings in relation to retention selectivity. J Chromatogr A 832:17–27
7. Shibukawa M (2006) Studies on the states of water in separation media of aqueous liquid chromatography in relation to separation selectivity. Bunseki Kagaku 55:149–162
8. Sillén LG, Martell AE (1964) Stability constants of metal-ion complexes. The Chemical Society, London
9. Suzuki N, Kodera J, Imura H (1987) Extraction equilibrium constants of scandium(III) and yttrium(III) in acetylacetone-organic solvent systems and some comparisons with rare earths(III). Inorg Chim Acta 128:261–266
10. Ohmine I, Tanaka T (1982) Salt effects on the phase transition of ionic gels. J Chem Phys 77:5725–5730

11. Livney YD, Ramon O, Kesselman E, Cogan U, Mizrahi S, Cohen Y (2001) Swelling of dextran gel and osmotic pressure of soluble dextran in the presence of salts. J Polym Sci B Polym Phys 39:2740–2750
12. Seidy-Esfahlan M, Khodapanah E, Tabatabaei-Nezhad SA, Salami-Kalajahi M (2023) Swelling and rheological behavior of preformed particle gel nanocomposite: simultaneous effect of pressure, temperature, and salinity. Polym Bull 80:8995–9013
13. Thiel J, Maurer G (1999) Swelling equilibrium of poly(acrylamide) gels in aqueous salt and polymer solutions. Fluid Phase Equilib 165:225–260
14. Nakamura S, Imura H, Suzuki N (1985) Solvent effect on the synergic extraction of thulium(III) with acetylacetone and 1,10-phenanthroline. Inorg Chim Acta 109:157–162
15. Nakamura S, Suzuki N (1986) Adduct formation constant in the synergic extraction of several rare earth elements with acetylacetone and 1,10-phenanthroline in nonpolar organic solvents. J Radioanal Nucl Chem 99:145–152
16. Petrova MA (2010) Adduct formation constants in the synergistic solvent extraction of several lanthanoids with two chelating extractants: β-diketone and bidentate nitrogen bases. J Chem Eng Data 55:5791–5796
17. Choppin GR (1981) Studies of the synergistic effect. Sep Sci Technol 16:1113–1126
18. Mathur JN (1983) Synergism of trivalent actinides and lanthanides. Solvent Extr Ion Exch 1:349–412
19. Saunders D, Pecsok RL (1968) Calculation of distribution coefficients in inorganic gel chromatography. Anal Chem 40:44–48
20. Neddermeyer PA, Rogers LB (1968) Gel filtration behaviour of inorganic salts. Anal Chem 40:755–762
21. Yoza N (1973) Gel chromatography of inorganic compounds. J Chromatogr 86:325–349
22. Tarutani T, Watanabe M (1973) Chromatographic behaviour of some divalent metal ions in sodium chloride, sulphate and nitrate solutions on a Sephadex G-10 column. J Chromatogr 75:169–172
23. Shibukawa M, Ohta N, Kuroda R (1981) Distribution processes of inorganic solutes in gel chromatography. Anal Chem 53:1620–1627
24. Shibukawa M, Ohta N (1986) Model for partition chromatography of ionic solutes in the presence of eluent electrolytes. Chromatographia 22:261–267
25. Marsden NVB (1985) Alkali cation selectivity of Sephadex G-25 in water and aqueous mixtures of methanol, ethanol and *tert.*-butanol. J Chromatogr 319:247–261
26. Fukuda T, Kohara N, Onogi Y, Inagaki H (1990) Separation of simple ions by gel chromatography: I. Simple model of separation for single-salt systems. J Chromatogr 511:59–68
27. Okada T, Miyakoshi M, Inoue M, Miyanabe M, Ueno Y, Nakabayashi Y (1992) Differential elution from a Sephadex G-15 column of sodium and phosphate ions of sodium phosphate with sodium or potassium phosphate buffer. J Chromatogr 604:219–224
28. Davankov V, Tsyurupa M (2005) Preparative frontal size-exclusion chromatography of mineral ions on neutral hypercrosslinked polystyrene. J Chromatogr A 1087:3–12
29. Davankov V, Tsyurupa M, Alexienko NN (2005) Selectivity in preparative separations of inorganic electrolytes by size-exclusion chromatography on hypercrosslinked polystyrene and microporous carbons. J Chromatogr A 1100:32–39
30. Laatikainen M, Sainio T, Davankov V, Tsyurupa M, Blinnikova Z, Paatero E (2007) Modeling of size-exclusion chromatography of electrolytes on non-ionic nanoporous adsorbents. J Chromatogr A 1149:245–253
31. Shibukawa M, Takazawa Y, Saitoh K (2007) Measurement of mobile-phase volume in reversed-phase liquid chromatography and evaluation of the composition of liquid layer formed by solvation of packing materials. Anal Chem 79:6279–6286
32. Shibukawa M, Kondo Y, Ogiyama Y, Osuga K, Saito S (2011) Interfacial water on hydrophobic surfaces recognized by ions and molecules. Phys Chem Chem Phys 13:15925–15935
33. Takeuchi T, Jiang B, Lim LW (2012) Separation of inorganic anions on hydrophobic stationary phases in ion chromatography. Anal Bioanal Chem 402:551–555

34. Nakamura K, Nakamura H, Saito S, Shibukawa M (2015) Surface-bubble-modulated liquid chromatography: a new approach for manipulation of chromatographic retention and investigation of solute distribution at water/hydrophobic interfaces. Anal Chem 87:1180–1187
35. Nakamura K, Saito S, Shibukawa M (2018) Adsorption at the water/hydrophobe interface versus partition into the interior of the hydrophobe: Quantitative evaluation of the solute retention selectivity at the water/hydrocarbon interface. J Phys Chem C 122:4409–4418
36. Nakamura K, Ubukata R, Mizuno H, Saito S, Shibukawa M (2018) Effect of acetonitrile on the solute distribution at the heterogeneous interface region between water and hydrocarbonaceous silica revealed by surface-bubble-modulated liquid chromatography. J Phys Chem C 122:28674–28683
37. Nakamura K, Saito S, Shibukawa M (2020) Intrinsic difference between phenyl hexyl- and octadecyl-bonded silicas in the solute retention selectivity in reversed-phase liquid chromatography with aqueous mobile phase. J Chromatogr A 1628:461450
38. Shibukawa M (2023) Surface-bubble-modulated liquid chromatography: an experimental strategy for identification of molecular processes of solute retention in reversed-phase separation systems. Anal Sci 39:791–813
39. Shibukawa M (1993) Dependence of retention of ionic solutes on the composition of the mobile phase electrolytes in partition chromatography. J Chromatogr A 655:199–205
40. Shibukawa M, Ohta N (1988) A new method for the determination of mobile phase volume in normal and reversed-phase liquid chromatography. Chromatographia 25:228–294
41. Kozaki D, Mori M, Nakatani N, Arai K, Masuno T, Koseki M, Itabashi H, Tanaka K (2013) Indirect UV detection–ion-exclusion/cation-exchange chromatography of common inorganic ions with sulfosalicylic acid eluent. Anal Sci 29:121–126
42. Foley RCL, Haddad PR (1986) Conductivity and indirect UV absorption detection of inorganic cations in non-suppressed ion chromatography using aromatic bases as eluents: I principles of operations. J Chromatogr A 366:13–26
43. Shibukawa M, Ohta N, Onda N (1990) Investigation of the states of water in water-swollen hydrogels by liquid chromatography and differential scanning calorimetry. Bull Chem Soc Jpn 63:3490–3494
44. Kaiser E, Rohrer JS, Jensen D (2001) Determination of trace anions in high-nitrate matrices by ion chromatography. J Chromatogr A 920:127–133
45. Dash K, Karunasagar D, Thangavel S, Chaurasia SC (2003) Procedure for determination of trace ions in boric acid by matrix volatilization–ion chromatography. J Chromatogr A 1002:137–144
46. Raessler M, Hilke I (2006) Ion-chromatographic determination of low concentrations of nitrate in solutions of high salinity. Microchim Acta 154:27–29
47. García-Robledo E, Corzo A, Papaspyrou S (2014) A fast and direct spectrophotometric method for the sequential determination of nitrate and nitrite at low concentrations in small volumes. Marine Chem 162:30–36
48. Wang S, Lin K, Chen N, Yuan D, Ma J (2016) Automated determination of nitrate plus nitrite in aqueous samples with flow injection analysis using vanadium (III) chloride as reductant. Talanta 146:744–748
49. Senra-Ferreiro S, Pena-Pereira F, Lavilla I, Bendicho C (2010) Griess micro-assay for the determination of nitrite by combining fibre optics-based cuvetteless UV-Vis micro-spectrophotometry with liquid-phase microextraction. Anal Chim Acta 668:195–200
50. Sun W, Zhang S, Liu H, Jin L, Kong J (1999) Electrocatalytic reduction of nitrite at a glassy carbon electrode surface modified with palladium(II)-substituted Keggin type heteropolytungstate. Anal Chim Acta 388:103–110
51. Pena-Pereira F, Lavilla I, Bendicho C (2009) Miniaturized preconcentration methods based on liquid-liquid extraction and their application in inorganic ultratrace analysis and speciation: a review. Spectrochim Acta B 64:1–15
52. Wang GF, Satake M, Horita K (1998) Spectrophotometric determination of nitrate and nitrite in water and some fruit samples using column preconcentration. Talanta 46:671–678
53. Karthikeyan S, He J, Palani S, Balasubramanian R (2009) Determination of total nitrogen in atmospheric wet and dry deposition samples. Talanta 77:979–984

54. Schaller D, Hilder EF, Haddad PR (2006) Monolithic stationary phases for fast ion chromatography and capillary electrochromatography of inorganic ions. J Sep Sci 29:1705–1719
55. Niedzielski P, Kurzyca I, Siepak J (2006) A new tool for inorganic nitrogen speciation study: simultaneous determination of ammonium ion, nitrite and nitrate by ion chromatography with post-column ammonium derivatization by Nessler reagent and diode-array detection in rain water samples. Anal Chim Acta 577:220–224
56. Kitamaki Y, Jin J-Y, Takeuchi T (2003) Simultaneous determination of inorganic nitrogen species by microcolumn ion chromatography. J Chromatogr A 1003:197–202
57. Heckenberg AL, Haddad PR (1985) Studies on sample preconcentration in ion chromatography: II. An automated, single pump preconcentration system for non-suppressed ion chromatography with conductivity detection. J Chromatogr A 330:95–111
58. Jackson PE, Haddad PR (1987) Studies on sample preconcentration in ion chromatography: V. Effect of the ion-exchange capacity of the concentrator column. J Chromatogr A 389:65–74
59. Huang Y, Pang J, Zhang S, Huang W (2024) Pretreatment methods in ion chromatography: a review. J Chromatogr A 1730:465162
60. Barron L, Nesterenko PN, Paull B (2006) Rapid on-line preconcentration and suppressed microbore ion chromatography of part per trillion levels of perchlorate in rainwater samples. Anal Chim Acta 567:127–134
61. Biesaga M, Schmidt N, Seubert A (2004) Coupled ion chromatography for the determination of chloride, phosphate and sulphate in concentrated nitric acid. J Chromatogr A 2004:195–200

The manufacturer's authorised representative in the EU is Springer Nature Customer Service Centre GmbH, Europaplatz 3, 69115 Heidelberg, Germany. If you have any concerns regarding our products, please contact ProductSafety@springernature.com

Printed and bound by CPI Group (UK) Ltd, Croydon, CR0 4YY

30/09/2026

02225308-0009